Acknowledg

My sincere gratitude must go to:

Frank Capon for his excellent cartoons (I hope we sell enough books to pay him!)

My business partner, Richard James, whose investment, encouragement and unflagging enthusiasm has enabled the whole Ideasun Limited venture to happen

My wife, Nandy, for putting up with me and the Ideasun project for all these years including supporting me at exhibitions

Dan and Lins, and Tim and May (and my delightful grand-daughters) for their faith in me

My friend, Charles Clark, for being on hand with valuable advice and respected opinions whenever needed

My young, gifted, design studio guy, Andrew Wills, and his artist contact, Paul Woolford, who came up with the cover graphics at very short notice

Copyright:

John Richards 2007

INTRODUCTION

> *Genius is 10% inspiration and 90% perspiration*
> **Thomas Alva Edison**

Rather like the penniless artist in his garret, the notion of an inventor slaving away in a workshop trying to come up with gadgets which might meet the desires of the population is a romantic one. It's both news material and story material. The media love it and are keen to devote space and time to it. I'm certainly prepared to go along with the press' stereotype and accept the description "inventor" if it gets me free publicity which may raise the company profile and result in sales.

Inventors have featured in books from Daedalus in Greek mythology, through Leonardo da Vinci during the Italian Renaissance, to the twentieth century when they also began to appear in films. Fictitious portrayals have varied from wacky characters like Roald Dahl's Caractacus Potts to fearsome ones such as Mary Shelley's Frankenstein. The genre has been further enriched by comical pseudo inventors like Heath Robinson and Roland Emmett.

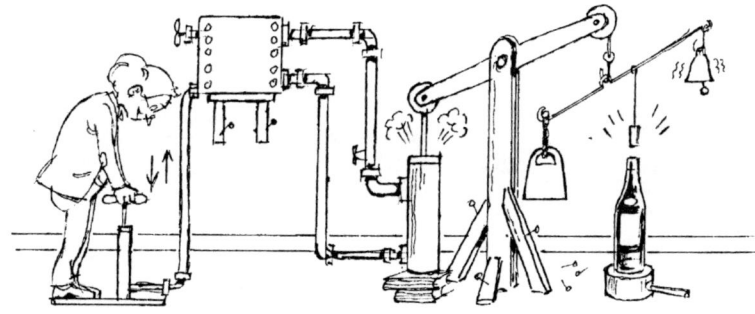

Biographical presentations such as the story of R.J. Mitchell, the designer of the World War Two Spitfire aircraft, as featured in the film 'The First of the Few', have added to the perception of the inventor/designer as a hero.

Why have I written this book? Well, I have learned a lot on my personal thirty year journey from concept to consumer and, as a former teacher, naturally I want to share my experiences **so that you might benefit** from my failures and successes.

> *He shows great originality which must be curbed at all times*
> **School report on Peter Ustinov**

Health Warning

It's tough being a lone inventor these days. Inventing used to be much easier. No, it's not because there is nothing left to invent; far from it. Simply, the conditions which favour the individual inventor were more prevalent in the past. The inventors' heyday was probably in the late eighteenth, the nineteenth and the early twentieth centuries when creative men were able to assemble the combination of time and money necessary to indulge in the development of their ideas. Nowadays, unless you marry wealth, it's difficult to escape having to pay the rent or the mortgage, which means you have to keep up the day job, so inventing has become more of a luxury pastime. Women joined in the inventing game a little later because, at the time of the famous historical inventors, not only was it mostly about heavy engineering akin to blacksmithing but, social attitudes required females to be conventional home bodies while men were given more freedom to pursue their fascinations.

Although there's still plenty left to invent, I do suspect we might be in the zone of diminishing returns for some fields of human endeavour. Take land transport for example: the big breakthrough was obviously the wheel. Centuries later the steam locomotive and the bicycle transformed society. These were followed by the automobile, which permitted luxurious personal mobility, and the tractor which industrialised farming. Since then we've had developments that penetrate the market much less: the armoured tank, the hovercraft, the quad bike, the Segway - vehicles that fulfil more narrow requirements. See what I mean? There's a peak followed by a falling off. You can track a similar story for many other classes of technological product e.g. writing implements. Exceptions are telecommunications and robotics which must have plenty of mileage still to go, but are they the domain of lone inventors?

Increasingly, creating new or better products has become a company endeavour; the prerogative of well funded teams of specialists. This is because working at the forefront of technology nowadays requires enormous investment in equipment capable of handling exotic materials at dimensions approaching the nanoscale in rooms that are clinically clean and environmentally controlled. Whereas Faraday could manage with everyday materials like iron magnets and copper wire insulated with cotton thread, today's cutting edge innovators make use of silicon crystals melted at high temperature and doped with a few atoms of phosphorus or boron to form microscopic components which are connected by gold. So, many of the bright ideas an inventor might have are beyond his capability to develop and he has no choice but to go, cap in hand, to the big boys.

Also, it used to be easier to do things that might be dangerous. We are so heavily regulated now that you can't even get hold of the ingredients for gunpowder legally, unless you are the bona fide chemistry teacher in a school or college. If the Health and Safety Executive had been around in 1813, William Murdoch wouldn't have been allowed to install gas lighting on Westminster Bridge because he wasn't CORGI registered!

Murdoch is a good example of an inventor; in 1777 he made a hat from wood (he called it 'timmer' – that was his pronunciation of 'timber') and walked three hundred miles to Birmingham to ask James Watt if he would employ him. You'd probably get killed on the roads trying that today. He got a job though; the wooden hat worked for him.

Then there's the problem of keeping people on board: wives, partners and family have to be persuaded that your project is not just a vain folly. There will be periods of economic hardship and marital strife. With the exception of a win on the lottery (statistically much more remote than being hit by lightning), big prizes can only be obtained by taking big risks. You will question your self belief; are you being single minded in a reasonable pursuit of investment, or selfish and conceited in your delusory gamble? Sometimes you are bound to wonder whether you are living in your own world like Walter Mitty or just spending time, in the words of the Buddy Holly song so memorably recorded by Dusty Springfield, "Wishing and hoping and waiting".

Being an inventor is a bit like being an outlaw. You don't fit into any of the pigeonholes of conventional society. All the nine-till-fivers will regard you with a mixture of awe and envy and they'll wonder how *you* got the right to pursue your own agenda while *they* have to conform to a boring job. Being employed is similar to being imprisoned in an office, shop or factory for forty hours a week. I used to be a secondary school teacher; that's like being incarcerated *with thirty spotty teenagers*. No wonder so many wage slaves yearn to dig an escape tunnel.

On top of that there's 'The Resistance', no, not the French Resistance, the 'Invention Resistance'. Believe it or not, the 'Establishment' doesn't want you to invent. They take a look at you, see you are flesh and blood like them, and can't believe you might have thought of anything new or better. There must be something wrong with your invention otherwise *they* would have thought of it. The Establishment has always done things one way; they know that way works, why should they change? They've seen iron sink, so they carry on making ships in wood.

It's worse if they know you. One of my old friends, who has a Doctorate in Zoology, used to have his mother argue with him over whether the animal they'd just seen was a rat or a mouse. In frustration he used to say she'd rather take advice from a blind beggar than believe him! This is due to an old human trait that was referred to in the Bible: a prophet is without honour in his own land. The fact is, it's always easier to impress people who don't know you or who only know of you by reputation. This may be why the more shrewd politicians don't reveal their policies until *after* they have been elected. These candidates recognise that voters can more easily support someone who, they are left to imagine since hope springs eternal, might share their own views. This effect is not just for the beginner; a leader is most likely to get booted out of *national* office when he is at his peak of *international* esteem. Mikhail Gorbachev, Margaret Thatcher and Tony Blair were all cases in point; just as other countries were admiring their statesmanship, the folks at home were getting fed up with their faults.

Next time I do a presentation I'm going to begin by asking if there is anyone in the audience who knows me personally and, if there is, I'm going to ask them to leave! Britain is probably the worst country in the world for this 'I'm not going to be impressed by anything home-grown' syndrome; countless inventors have had to go abroad to get backing. Sadly, 'The Resistance' is bigger and worse today than ever before; so, would-be inventor, you'd better be prepared to become a traveller.

Finally, please notice that this book is made of paper and don't regard the text as written in stone; if there's one rule I've learned about getting a concept to the consumers it is that there are no rules. Only judgements...

> *If you try to change things then you upset the establishment which is why invention and vilification have always gone hand in hand.*
>
> **James Dyson**

So, why do you want to invent?

It's about getting rich, isn't it? Be honest, most of us want a shortcut to wealth that involves minimal effort. We go about our drab and wretched lives as wage slaves, working mainly for the benefit of the mortgage lender or the landlord, dreaming about instant fame and fortune. We've heard about the Cyclone vacuum cleaner man, James Dyson, and Catseye inventor, Percy Shaw, and think we can do the same.

> In 1933 Percy Shaw was driving to Bradford in dense fog. Seeing his headlights reflected in the eyes of a cat sitting on a fence, he was able to stop just before driving over a precipice. Percy set up Reflecting Roadstuds Ltd to manufacture his "Catseye road reflector" and in 1937 it won a Ministry of Transport competition to identify the best road reflector. "Catseye" is now a household name throughout the UK.

Firstly, what is an 'inventor'? The online Oxford dictionary says, "Someone who creates or designs a new device or process". It's a strange coupling of concepts; to create is much more innovative than to design, surely? I could design a chair but I wouldn't claim to have invented the chair. I suppose the dictionary compilers are trying to focus on the *idea*, which doesn't have to be made into a physical form to be a patentable invention. However, the really operative word in the dictionary definition is 'new'. We are immediately into the area of the legal protection of intellectual property, about which more later.

The premier attraction of inventing is due to the fact that it may result in a *product*. Selling products is potentially so much more profitable than the other type of business: providing services. Obviously, if you own a huge company like Virgin airline, employing hundreds of staff to provide your service for you, then you can become enormously rich. At that level your highly trained staff are like your company's product, but if you are a one-man service industry, such as an architect, lawyer, bricklayer or plumber, your income is limited by the number of hours you can work. Whether you charge £4 or £4000 per hour for your services, you can comfortably deliver a maximum of only about 60 hours a week. Machines, however, can work 24/7 (168 hours a week) churning out a product every few seconds, and if you generate more demand, you simply get another machine. There is no limit! (You can even do both, of course; sell your products *and* command a fee for going on the lecture circuit telling others how to do it!)

The second appeal to inventing is that it might get you exclusive rights to a new or better product which has *no competitors*. If you get your invention protected you may have twenty years of monopoly in the marketplace as sole manufacturer, or control of the rights to manufacturing if you choose to go down the licensing route, or both manufacturing *and* licensing if you want a mixture, maybe in different territories. So, having control of a patented product is the way an inventor might hope to get really wealthy. And never forget that a business is also a piece of property; some of the richest men and women make a habit of building a business up for three to five years and then selling it at a profit so they can use the proceeds to start again. It's called 'having an exit strategy.'

Finally, an inventor has the pleasure of pursuing his own project rather than having to do the bidding of an employer. Inventors have their own agenda and they want to execute it more single-mindedly than most people. I realise that all self-employed people get to run their individual lives but this is something else. Your local greengrocer may be his own boss but he knows that every day he is going to be shifting boxes of fruit and vegetables. Inventing promises a much more interesting life.

Inventing promises a much more interesting life!

Creativity demands that you leave your comfort zone, continually challenge yourself and be prepared to confront conventional wisdom. Especially, engage in that for which you have not been schooled.

**Allan Snyder,
Director of the Centre for the Mind, Australian National University**

What do you need to be an inventor?

You need ideas. We all have ideas, in fact Trevor Bayliss (the clockwork radio inventor) says there's an inventor in all of us. My definition of an inventor is someone who sees what everyone else has seen and then thinks what no-one else has thought. So a spark of originality or ingenuity is required but this is the easy bit that most people can do. True, some will do it with

> **A Faint Grasp of Reality**
>
> A very rich man once asked me why I didn't invent a car that runs on water. He obviously saw water as just another liquid, like petrol, that you can pour into a tank; the fact that water is an end product of combustion, and therefore not a fuel, seemed to have passed him by.
>
> **The Author**

more understanding and, as a result, their output might be more relevant. Obviously you have to have a faint grasp of how things work in order to be able to suggest sensible new solutions. It was scientific ignorance that allowed the alchemists of olden times to waste their lives on trying to turn base metals into gold or discovering the 'universal solvent'. (What would they have kept it in?) However, **to foster creativity it pays to be open minded and interested in everything.** (This makes it sound too easy so I should warn you here and now that the real difficulties come much later.)

Being open minded enables you to 'think outside the box'; to imagine alternatives to the existing solutions or to import some piece of knowledge from an entirely different field and

> Problems cannot be solved by thinking within the framework in which the problems were created
>
> **Albert Einstein**

apply it to the problem in question. For an inventor, counter-intuitive is good. It's also an attitude of cussedness; refusal to meekly accept boring things like conventional wisdom or rules. Success often comes from sidestepping the predictable route: the Arctic Monkeys got hit records from sales off their website promoted by the text messaging of their fans, rather than through the traditional music business processes. Robin Bines made a similar smart move to get Ecover detergent in front of UK shoppers (see later) and what about the no-frills airlines or Apple's iTunes? Sometimes the old mould just has to be smashed. Inventors need to be moduloclasts: 'mould-breakers'. OK, I made that word up. :-)

Meredith Thring (former Professor of Mechanical Engineering, Queen Mary College, University of London) referred to this outlook as the 'Inventor's eye'. He said an inventor must look at every human device as if he were a man from Mars and ask himself, "Why do they want it that way?" This type of mind-set includes what has become known as 'lateral' thinking and 'joined up' thinking. The classic example of this was James Watt who, as the story goes, wanted to find a better source of power than the horse or the windmill, when he noticed the lid of the boiling kettle lifting…

People who are interested in everything have more chance of making a valuable connection that no-one else has spotted. Maybe I'm talking about 'geeks' like me here but, speaking personally, information is my entertainment. I'd rather see a television programme on, say, wind turbines or artificial hearts than watch a drama. The experience, even if vicarious, of different technological solutions gives you an armoury of useful information which may have other applications one day.

Embarking on developing a product from your idea is an adventure so you also have to be prepared to take a risk. We humans, as a species, have evolved to be

> "If I had to live my life again, I'd make the same mistakes, only sooner."
> **Anon**

naturally conservative. The more daring early hominids approached an unfamiliar creature too trustingly and didn't live long enough to pass on their genes to many offspring. Sticking with the known is safer than changing for something new. Unfortunately, conservatism is the enemy of invention and it's rife here in Britain. Too many UK people hate change. Inventors have had to fight against official disinterest, opposition and lethargy to get investment in their projects; just think of the red flag law. The USA is much more encouraging; inventing is part of 'The American Dream'.

So, if you are reading this on an aircraft you are possibly the right sort of person to pursue an invention to a conclusion; but, if you are sat in a comfortable armchair by the fire with your dog curled up at your feet and a pipe of tobacco in your mouth, then you might as well give up any inventing ambitions. I've never heard of a successful 'armchair inventor' but I do think that inventors and explorers must have common characteristics such as curiosity, initiative, boldness and determination.

> **Red Flag Law**
>
> The law in Britain, until November 1896, required motor vehicles to be preceded by a pedestrian carrying a red flag as a warning to other road users. This effectively restricted the vehicles' speed to 4 mph.

The reason you have to be a risk taker is because all business is a gamble. You are going to have to back your ideas and opinions with your own money. You may be able to get financial help but, even so, you will be required to contribute a stake yourself if only to demonstrate your faith in, and commitment to, the project.

If you doubt whether business is a gamble take a look at the case of IBM. International Business Machines was once the major computer manufacturer on the planet but now they don't produce a single laptop or desktop. Their personal computing department, which more or less invented the PC in 1981, was acquired by the Chinese firm, Lenovo, in May 2005.

> **Chinese proverb**
>
> The worst mistake you can make is to live in fear of making one.

Back in the early eighties, IBM gave away the job of writing the software for their new personal computers to a little known Seattle company called 'Microsoft' led by a young nerd name of Bill Gates. Yes, I did say *gave away*. They gambled that there wouldn't be any money in selling the software! Having given away the software part of their business and lost the hardware PC production part, what are they doing now? Please don't laugh, they're offering a business consultancy service!

What are these difficulties that I've hinted at earlier? How about: finding time, trying to tolerate delays in obtaining supplies and estimates, paying fees and bills, for a start. Money problems essentially, for most of these things could be solved by a bottomless pocket. Just as an example, you won't believe how difficult it might be to locate a seemingly simple thing like a source of ball pen refills these days. Nobody seems to make them in Western Europe anymore. They all want to sell you complete pens. Manufacturers of refills only sell to pen makers and they won't tell you who their suppliers are. That's just a tiny sample of the frustrations you are going to encounter, and to cope with them **you will need enormous reserves of tenacity.** This is probably what Edison meant when he said, 'Genius is 10% inspiration and 90% perspiration'.

> We choose to go to the Moon in this decade and do the other things, not because they are easy, but because they are hard.
>
> **John F Kennedy 1962**

However, difficulties in development and production are as nothing compared to the biggest problem of all: marketing. You have been warned! Read the James Dyson story "Against the Odds" if you want to know the full extent of the determination you are going to need. You remember on page eight I suggested that the reason we dream about inventing is because we want a shortcut to wealth involving *minimal effort?* Well, you can forget that idea.

Eric Laithwaite, (late Professor of Heavy Electrical Engineering at Imperial College of Science and Technology, London) who invented the linear electric motor was, like me, an only child and he thought that had an effect on his inventive capabilities. The evidence is to the contrary; inventors have come from various sizes of families: Edison was one of seven children.

I do wonder whether Prof Laithwaite had a point though because, not having been under pressure to play with a brother or sister, I was always free to indulge in the pleasure of my own thoughts and activities as a boy. This probably made me content with my own company and I wasn't lonely since I had readily available playmates in the form of cousins or school friends living next door throughout my childhood. Conversely, as a person without blood ties to my peer group, I never had the luxury of feeling that I could rely upon the loyalty of a playmate. When they are children, siblings can afford to ill-treat each other, safe in the knowledge that they'll be sleeping on the same bunk beds come nightfall. Without that certainty of companionship, I soon comprehended the need to be good company and the desirability of harmonious relationships. Maybe this enabled me to develop some skill as a peacemaker and humorist, he said modestly.

As a grammar schoolboy I was given the genial nickname 'Prof' because of my tendency to design weird gadgets and indulge in flights of fancy and, when I was teaching much later in life, this epithet was again attached to me by the boys at a Special Needs school. (I ought to confess that I have also been called much worse things.) **It certainly helps to be personable** because an inventor needs to be able to convince potential investors that he is worth backing.

Another factor that you need is luck. I don't mean rabbit's feet and horseshoes; I don't believe in superstition. The kind of luck an inventor needs is the sort we call 'serendipity'. It's a case of spotting when something fortuitous happens to you and taking maximum advantage of it; I give an example of this happening to me in the section on prototype production. However, there are some ways you can influence your own luck. That is by putting yourself into a position where the right things can happen to you. It's a bit like those card games where the first thing you do, after the deal, is pass on a couple of cards from your hand to the next player.

> *Accept nothing at face value and get in the habit of thinking unconventionally. Work hard, work smart and, with a bit of luck, serendipity will play its part.*
>
> **Australian Nobel Prize-winner Peter Doherty**

If you short suit yourself the chances are, after a round or two of harmless tricks, you will be able to unleash your poison onto the other players. When you are playing Monopoly, my advice is to try to build hotels on the orange properties. There are things similar to those that you can do to improve your opportunities in all aspects of business, including inventing new products.

To sum up then, an inventor needs to be creative, open minded, interested in everything, adventurer/risk-taker/gambler, tenacious, personable and lucky. Not much to ask for! Oh yes, 'rich' would help too. Some inventors in the past took the precaution of being born to rich parents to address that need.

> *When all has come right, the kind of man who persisted despite constant ridicule from the controlling forces will be said to have possessed vision.*
>
> **James Dyson**
> **in**
> **'Against the Odds'**

Are you capable of inventing?

Yes. You are. Let's face it, birds, sea otters and chimpanzees have invented tools to help them get food and you're at least as intelligent as them! In case you are not familiar with those examples, here are the stories.

The bird in question is Hunt's New Caledonian crow which has been observed, in its native group of islands 900 miles Northeast of Australia, using two distinctly different kinds of tools to forage for insects, centipedes and larvae. One tool is hooked, which the birds make by plucking and stripping a barbed twig; the other is serrated like a saw and is made from a leaf. Animal behaviourist, Dr Alex Weir says, *"My most striking finding was that one of our birds, 'Betty', spontaneously made hooks to retrieve food by bending pieces of wire. That is something even our closest relatives, chimpanzees, seem incapable of."*

The sea otters off the California coast forage close to the shore, diving in kelp forests, along rocky reefs and over the muddy bottom. They pick up urchins and crabs or pry abalone and scallops from the rocks using a stone as a tool. Then they eat while floating on their backs using a rock as an anvil to crack open hard-shelled prey. The otter places a flat rock on its belly and bashes snails, mussels or other tough prey against it. That's pretty inventive.

Chimpanzees in the Congo Basin regularly visit two different forms of termite nests and they use specific sets of tools to extract their insect prey depending on the structure of the nest.

The other day I watched a pigeon shortening a twig to make it suitable for the place she had in mind for it in her nest building enterprise. Even wasps have been observed using tiny little sticks to tamp down soft mud around their nest. You should definitely perform somewhat better!

Anyone can become an inventor as long as they keep an open and inquiring mind and never overlook the possible significance of an accident or apparent failure.

Patsy Sherman, inventor of 'Scotchgard' fabric protection coating

---

There are no failures, only feedback.

Tycoon Peter Jones

Can you develop your inventing skills?

Another yes! If you study the lives of the great inventors you will find that most of them 'messed about' in kitchens or makeshift workshops learning about the properties of materials and methods of working with them. Don't just immerse yourself in the history of invention; keep up to date with modern developments in any field that interests you. I subscribe to New Scientist – it's a serious source of news and information about everything scientific or technological and not all 'Gee Whiz froth' like the American magazine, Popular Science.

> *We advance when smart people try to solve tough problems.*
> **John Mankins, NASA's Advanced Concepts Program manager**

I was once asked by an insightful radio interviewer whether I had a Meccano set as a boy. The answer, of course, was, *"Yes."* I also did woodwork at school and learned about motor mechanics from my late father who was an erstwhile toolmaker in a factory which built parts for the famous World War Two high speed, twin-engined Mosquito aircraft. After the war he repaired and serviced his fleet of hire cars and taxis in his own workshop and we did things together that mechanics wouldn't attempt today, like rebuild gearboxes. Just watching him using tools was an education. I must admit it was easier then; mechanisms were deconstructable using a spanner and screwdriver whereas nowadays they come in sealed boxes with notices proclaiming *'No user serviceable parts inside'.* In my young day a more appropriate notice would have been, *'May contain nuts'* (and bolts)!

Also, Britain having recently been on the winning side of a war in which science and technology had played a big part, those subjects were heroically fashionable and interested boys like me did chemistry at home as a hobby. Naturally, in the nineteen fifties, we all had fun making explosives after a trip to the chemist, grocer and hardware store for the then readily available ingredients! How lucky we were not to have been born later when the stranglehold of health and safety regulations began to get a grip.

Another thing you can do is take every opportunity to investigate machinery. Go to the heritage sites when they have a 'steam up' and watch the old engines operating. Take things apart and try to put them together again. Get used to drilling, routing, riveting, turning, threading, etc. Find out what adhesive is good for which materials. Learn about vacuum forming, injection moulding, extruding, sputtering, etc. Do you know what a 'jig' is, or a 'template', and what about a 'die' or a 'ratchet'? How about a knurled grommet? Ok, I made that one up. :-)

WARNING:

Don't take things apart until their useful life is over and certainly not if they are plugged in.

Particularly beware of old televisions which have a large capacitor (next to the mains dropper). Just pointing your finger at it can give a nasty shock even after it has been switched off and unplugged.

The more you know about **different materials and how to process them**, the better you will **be** able to come up with a solution to a problem. It's simple; the more you play about with things, the more you are likely to invent something, even by accident! You'll know when you've succeeded because people will say to you, *"Why haven't we been doing it like that before?"* or the ultimate accolade, *"I wish I'd thought of that!"*

In my late twenties I bought a derelict old cottage and plumbed it, wired it, designed and built a garage, an extension and a loft conversion, built brick and flint walls, timber and slate roofs, did plastering, carpentry, etc. Amateurs are discouraged from doing much of this these days if only because, when you sell a property, the purchaser's solicitor now asks for all the guarantee certificates. My, admittedly conceited, attitude was: I have a brain and two hands, the same as any tradesman, so I should be able to do these tasks. More recently I have become quite useful at using a computer. Developing skills like these enhances what Professor Thring referred to as 'thinking with the hands'. He probably meant developing the flexibility of the brain to accommodate new capabilities.

Professor (Lord) Robert Winston, in his recent TV series, describes 'learned helplessness': the fixation that some people adopt that enables them to say proudly, *"I can't do maths"* or *"I can't sing"*. He says they have just *chosen* this attitude. A mindset that is content to accept and acknowledge, or even boast about, an inadequacy is less likely to invent. Inventing is not going to happen unless you are a self starter and determined to attempt anything. Anyone whose ambition is limited to waiting to being told what to do by a boss might as well stop reading now.

> *There was, and I believe still is, a pride taken by academics in **not** knowing how to make things with their hands, although why ignorance of anything should be cause for celebration I really don't know.*
>
> **James Dyson**

James Dyson wrote the quote above before the advent of reality TV star Jade Goody who seems to have become a celebrity for being ignorant!

FROM CONCEPT TO CONSUMER

What should you invent?

You should aim to invent **the ideal product**. What is that? Well, it is the product that most nearly meets the following ten criteria:

   1. **Invent something people *want*** (not something they need!)

This is not as obvious as it sounds. What do I mean? Well, the more your invention resonates with the population's sense of values, the more money you will make. Sadly, there's not much profit in inventing something *worthy*. Some folks I met at an Education Show brought this home to me. This was a *failure* of a show. There was only a handful of visitors and the organisers, who had taken thousands of pounds off us exhibitors on the pretext that we would have crowds to present our wares to, stayed cowering in their office to avoid our wrath. However, they had invited some private schools to exhibit on the theory that they might attract new pupils from the 'hoards' of educationally minded parents that the organisers alleged would visit the Show in this rich part of Surrey. These schools became my customers because I abandoned my own stand and went round selling to them. On my tour of the Show I chatted to the other stand holders and one couple were franchisees for a small chain of educational toy shops. It was a little rival to the Early Learning Centre. (The ELC started out as a showcase for learning equipment and then turned into a toy shop after it had been taken over; it's been taken over again but it still focuses on toys and now they are almost exclusively own-brand.)

The franchisees told me about **'Rosemary's Brussels'**. It transpired that their franchiser, Rosemary, was an ex-teacher who kept supplying them with products that she deemed would be good for helping children to learn. On her visits the franchisees dutifully put her latest stock on the shelves then, when she was gone, they took it down and replaced it with teddies, crayons and balloons, etc. Why? Because her stuff didn't sell. The children's parents, who had the purchasing power, wanted to buy their offspring something to give them *pleasure* not something associated with the 'work' of learning! Hence 'Rosemary's Brussels': we all know that Brussels sprouts are good for us, but nobody likes them! This little chain of shops failed to respond to public demand in the way that the Early Learning Centre had and soon had a financial crisis. Luckily an investor came along with a rescue package but, unfortunately, the business world only teaches lessons the hard way; I guess the franchisees lost at least some of their investment and I know the franchiser lost a piece of her company.

(Frank drew cabbages because Brussels sprouts are too small!)

Isn't it a pity that more adults don't understand how much pleasure children get from learning. Young children have brains designed to be curious and to avidly absorb information, which explains why they are always asking questions, but many new parents can only relate back to their own attitudes as surly teenagers in secondary school, so that's what they assume learning is like for their children. It's anthropomorphism applied to our own species! Sorry, I'm digressing.

The other side of the coin is shown by diet books and exercise machines. Despite the fact that you need to run 100 miles to burn off the calories in one chocolate biscuit (ok, that's an exaggeration, it's actually walk four miles), the public has an inexhaustible appetite for side-steppers, ab-trimmers, etc. Several television channels get employment marketing them. Almost certainly they won't actually get used for long but, as long as they are bought, the manufacturer will be happy! **The message is simple: any vanity based product has a good chance of success. Aim to gratify your customers or indulge them!**

We humans are all hypocrites. We may *say* that nurses, teachers and firemen are the most valuable members of society but we pay them such poor wages they can't afford to buy a house. When we *do* open our wallets who do we pay well? Footballers and pop singers! It's the same with products. The largest proportion of our spending is on non-essential items for our own gratification: cosmetics, confectionery, hairdressing, fashion garments, jewellery, tattoos, gas guzzling sports utility vehicles, computer games, tickets for football matches or other entertainments, alcohol, etc. Microsoft seem to have the right idea; they have just invested (gambled) billions of dollars on their second generation game console/home entertainment centre which they see as the next 'must have' for the hedonistic population.

There is a *need* for gadgets to make life easier for the disabled, or to solve problems in the developing world, but these are cash-poor or small markets so there's little profit in meeting their *needs*. This harsh truth is exemplified by the pharmaceutical industry. Malaria kills an African child every 30 seconds and is responsible for more than 1 million deaths a year. There is little doubt that, if some serious money was thrown at the problem, a vaccine or cure could soon be found but where's the commercial value in that? Most malaria victims are struggling to exist on $2 a day so, instead, millions of dollars are being spent on research to find cures for baldness and premature ejaculation because the sufferers from those 'conditions' can afford to pay for a 'treatment'. There's much more profit obtainable in the pharmaceuticalisation of the good health of wealthy folks. I'm reminded of a gag by Tom Lehrer, the Harvard mathematician and pianist/comedy song writer who achieved fame in the sixties. In between numbers he tells us about his friend who is a Doctor. The guy decides to specialise: *in diseases of the rich!*

> *The quest is on for the female equivalent of Viagra and entry to a business allegedly worth more than £2 billion.*
>
> **Sunday Observer 12 June 2005**

Bill Gates deserves a mention here because, in 2003, he and his wife Melinda donated $168 million from their personal fortune for malaria research and they did it again in 2005 – this time $258m. Altogether they have poured more than $6 billion into global health; what the market won't do is left to the benefactors.

A serious problem comes from the fact that people may not know what they want. Henry Ford once said, "If I'd asked people what they wanted they would have said, 'a faster horse'." However, if you start by recognising the vain and shallow nature of your fellow human beings then you will stand a much better chance of making a product that sells well.

> *'Vain and shallow nature'? That's a bit harsh! How about 'enjoyment oriented'?*
>
> **The Author**

This is advice that I am only just beginning to accept myself; having spent most of my life working in the public sector, out and out capitalism has taken a lot of coming to terms with. The few black Africans in South Africa, who have become wealthy since Apartheid was overthrown ten years ago, are ahead of me with this problem. Some of them are former African National Congress bosses who have completed the journey to riches. Now they are experiencing alienation from their fellow black South Africans (and from whites everywhere) who were under the impression that these leaders had been seeking justice and equality, not a personal fortune. From their viewpoint, the newly affluent ANC men can't understand why all those people, who famously championed capitalism themselves in the past, won't accept and admire *them* today just because they have only recently embraced it.

The pendulum swings; there is no happy medium, so you have to overcome your moral objections and invent something to meet a *want*, like a better hairbrush or an exercise machine that purports to reduce weight. If you do become rich, then you will be able to become a philanthropist and salve your conscience.

Brushlite: the wonderful weight reducing hairbrush!

2. Don't invent something too similar to an existing product.

Why should people buy your slightly better corkscrew? Can you imagine the uphill battle you will have convincing folks, who already have a satisfactory corkscrew, that your new design is worth paying for? The product will crawl off the shelves unless it is genuinely better, cheaper, more trendy or, preferably, all three. So look for a gap in the market.

Take board games as an example. Every year new board games come on the market at Christmas. They sell a few and then are forgotten, while Monopoly and Trivial Pursuit soldier on, constantly being reinvented with new variations. **If a marketplace is already occupied by strong products, you will have little chance of invading it successfully.**

I saw a TV programme recently about two guys who started a couple of burger bars in a city. They claimed their burgers were 'real' and imagined that would be enough to persuade the public to buy them. One of their outlets was across the road from a branch of Burger King. How they got it into their heads that they could take on this giant and beat it, I just don't know. Even if their burgers were 'better', when the customers got it into their mouths it wouldn't be the taste they were *familiar* with. Haven't those guys tried getting children to venture into eating a slightly different food? They must like a difficult job.

Even if 'Proper Burgers' got a massive level of trade, the competitor across the street is so rich (rich = powerful) they could easily beat their upstart rival in a price war. Burger King would only have to undercut their price for a week or two and watch them go bust. Business is about survival of the fittest. Why try to reinvent the burger? Steer clear of things that have been done like chicken nuggets or pizzas. What's wrong with starting the definitive Welsh rarebit and tea bar? Have you tried mixing grated apple with grated cheddar and grilling it on toasted crumpets? It's all fluffy and wonderful, especially with a little freshly ground black pepper. You'll find it's always better to be in an uncompeted field and inventing something new gives you the opportunity to do that (but see ideal product criterion number three).

You must beware of falling into the trap of believing you can make a business work just because you want to get out of the day job. You may believe you can earn a fortune selling English muffins covered in melted cheese drizzled with Worcester sauce but *belief* is not significant. Belief is just a personal choice and can change. I used to believe I liked sweetened coffee; now one crystal of sugar ruins it for me. What *is* important is *evidence*. Do some research. Find out if there is a market for hot cheesy crumpets (see Trial Marketing). Collect the evidence – it may represent a commercial reality, but don't forget Henry Ford's words.

A word of warning about market research would be timely here though: people tell researchers what they think they want to hear, or what they want to believe about themselves. The American Burger restaurant chain found this out when they added a fresh-fruit bowl to their menu. One year later they quietly killed it blaming lack of demand. "We listened to consumers who said they wanted to eat fresh fruit," a disarmingly honest spokesman told the New York Times, "but apparently they lied." (The Guardian 23 08 06) So market research is neither infallible nor essential: see more about this under the section called Market Research and Trial Marketing.

> *My principal business consists of giving commercial value to the brilliant, but misdirected, ideas of others.... Accordingly, I never pick up an item without thinking of how I might improve it.*
>
> **Thomas Edison**

However, if your product is a short-lived disposable or consumable item you can forget this piece of advice about not making something too similar to existing products. There will be a constant demand for replacements and, assuming your product is an improvement, cheaper or in some way fashionable, its time will come when customers are seeking to renew their existing stocks. Ball pens, socks, toilet rolls, Christmas baubles, candles and vending machine cups are the sorts of things that come into the consumable class (please don't mention a singing fish on a wall plaque).

After all, Ben and Jerry did it. They saw a marketing opportunity for an expensive luxury ice-cream and now you can buy it everywhere; I bought some in Accra, Ghana, West Africa, only last week. It's probably no coincidence that their product is in the 'indulgence and self gratification' category of goods.

Slightly different corkscrews

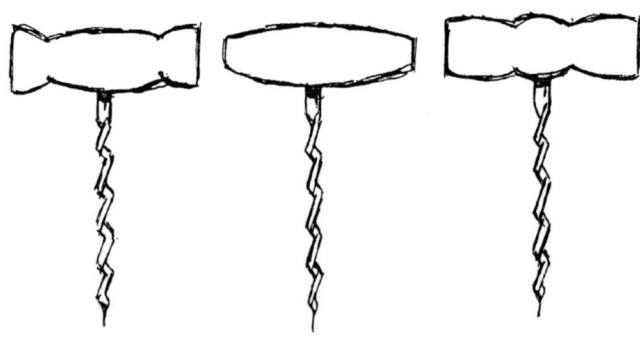

3. Don't invent anything *too* new or different.

If it's too much of a leap of understanding for the potential consumer to grasp they won't buy it. You see, the trouble with some inventions is, they're *too new*. 'New' means 'risky' to a surprising number of conservative shoppers unless it's just a meaningless claim made on a well-known product like detergent. Rather than buy something genuinely new they'd prefer to spend money on something *comfortably familiar*, like a book.

I've been at lots of Education Shows selling my really useful innovative teaching aides and I've watched folks take hours to browse round the bookstall or sticker stand opposite then, when they've finished that, they just steam straight past my little stand without stopping for a glance. It's as if they feel guilty about spending all that time at the bookstall and want to get the hell out of there. I could be giving away the elixir of life and they wouldn't even know. They've done the *comfortably familiar* (books) and now they are heading for the *mundane* (stickers). Actually they may have come specifically for books or just suddenly need the toilet, but I'm beginning to feel somewhat neglected and slightly paranoid.

Any really different new product involves teaching the potential customers what it can do before they will realize that they need it. As a former teacher I can tell you that this will be very difficult to achieve, and therefore enormously expensive. It's tantamount to social engineering and there are few more challenging tasks to attempt. It's easier in the States where people are more receptive, but sceptical UK purchasers do not welcome being lectured to. At that time our customers were mostly female teachers who are used to being regarded as authorities themselves and they often reach for the word 'patronising' if they think they are being told anything, especially if the presenter is *a man*. If your product involves too new a concept, at best there is going to be a time delay while your costly marketing strategy brings potential customers up to speed, at worst they may never cotton on. If you can't get the benefits across in 30 seconds you probably won't be able to sell your product (see The Pitch).

We had difficulty with the Wordwiza reading tool in this respect. It has to be demonstrated in order to sell. Show people a static picture of it and their reaction is, *"What's that?"* There was nothing like it before. Imagine how difficult it must have been to sell the first ruler. A ruler is just a thin piece of wood or plastic until you know what it's for. We sold thousands of Wordwiza from our stands at the Education Exhibitions where teachers could be shown that it enabled children to dissect words with their hands, but nobody would buy it from a catalogue. It took a flash demo on our website (www.ideasun.com) to overcome that problem.

It's hard to get across just how difficult it is to educate the public. Take this example: a friend of mine wholesales gift wrapping paper and he got a new supply from Germany. The new paper was twice the size of the previous product and it wouldn't sell in the UK. Despite describing it as super sized and special value, British people still preferred the stuff on rolls or the original smaller folded packs. Don't ask me why they couldn't make the adjustment to a bigger sheet – I simply don't know.

When I was on a solar boating weekend in Switzerland a few years ago I met a young man who had come up with the most amazing invention. It was like two chains that you pushed through a gate which operated like the runner on a zip. The chains went in all floppy but came out the other side interlinked into a rigid rod. It was utterly brilliant! When you pushed the rod back through the gate it became two floppy chains once more. He'd had it made in both plastic and metal. It's a sort of technical solution to the Indian rope trick. Unfortunately none of us has been able to suggest a sensible application. What's it for? Is there a tent manufacturer reading this who would like collapsible tent poles? Would the army like to be able to extrude bridges across chasms and then reel them in and take them away?

Just remember, it's much easier to tap into the public's knowledge of existing products and sell them, for example, a better pen or vacuum cleaner. I know this sounds as though it contradicts criterion number 2 but, somewhere in between the extremes of too similar and too new is where the ideal products lie; they have enough innovation but not too much.

There's no mercy out there.

John de Groot
Sussex Enterprise

4. Maximum appeal.

You want to produce something which is as ubiquitously desirable as possible. Birch twigs may go well in Finland where, rumour has it, they enjoy flagellating themselves in the snow after a hot bath, but they are not popular elsewhere and, let's face it, Finland is a small market (population 5.25 million).

Barbie dolls may be big with girls aged between 3 and 6 but what about all those boys, teenagers and adults? How many younger folk will buy a pipe and carpet slippers?

"Why not? They're lovely!"

You see what I mean?

The most ideal product will appeal to both sexes, all ages, and all nationalities.

A high quality appearance is of paramount importance in getting maximum appeal – obtaining a sale may be more about the packaging, the advertising or the point of sale presentation than about the product itself. Many times I have seen folks presenting their products and thought to myself, *"I'm glad I don't have to try to sell that!"* Although it can be hard to tell which products will be successful; a diary and pencil is much better for most purposes than a Personal Digital Assistant but they sell enough of those. (I did meet a firm of traitors at the BETT 2006 Show. They sold PDAs but their company policy was for them to use a pen and notepad because, yes, it's better!)

You can't sell products on appearance alone though; except porn of course! One stand at an Education Show was trying to sell a beautifully manufactured board, with coloured magnetic sliders on it, which purported to be a wonderful profiling assessment tool. It was no surprise to me that they didn't sell any. They were targeting a very small market, it took too long to explain the benefits, and you could replace it with a printed form. In fact you would need to record the final position of the sliders on paper anyway. However, if someone was to do a software version...

One of my early ideas was for a 'Bunsen burner storage rack'. Storing Bunsen burners is a nightmare; a drawer is no good because it just needs one burner to settle on its base with the brass tube sticking up inside the frame of the unit and the drawer won't open; a cupboard is no good because the rubber hoses get into a bird's nest of a tangle then one springs out through the doorway and stops the door shutting. My racks worked a treat, keeping everything tidy and to hand. They were based on a piece of plastic drainpipe and were very space-saving. My father developed a machine to produce them based on four electric pillar drills operated by a foot pedal. The trouble was the size of the market. England and Wales have a total of about 4000 secondary schools, every one with maybe 3 or 4 science laboratories which perhaps could each take 2 racks: 32,000 units at the most. Even given an incredible level of market penetration, such as 10%, that's only about 3,000 sales; minimum appeal like that is not worth tooling up for. **You need to do a similar calculation before re-mortgaging your house to finance *your* invention...**

I recently had an email from a helpful customer suggesting that we should make an electronic reader for exam candidates who have the reading concession. He wrote, "10% of candidates have some degree of dyslexia, so that's quite a market". No it isn't! For a start it's just a few cohorts of the total population. Then, you might only penetrate 5% of that 10% - 0.5% overall. It's the other 90% of examinees that is 'quite a market' and aren't we forgetting all the other age groups of people that don't do exams?

It is deceptive: about one million people sit written exams every year in England and Wales so it sounds like a lot but, the 10% with reading problems is one hundred thousand and a wild success would be to sell to 5% of them which is just five thousand. That is a generous estimate of the maximum you could sell. Would you invest thousands in the necessary research and development and the costly tooling up to manufacture a product for that size of market? How would you promote it? Don't say, "But it would sell worldwide!" Can you speak all the other languages? What about the exchange rates? Import Duties? Shipping costs? **You know your own country best; it's your easiest market so try to succeed at home first.** Do you have an existing, cost effective, route to the consumer? If not, leave it to someone who has, because you won't be competitive.

Never waste time inventing things that people would not want to buy

Thomas Edison

5. Low development costs.

You may not have the millions to invest in developing a complex product that some of your competitors have. If you are an inventor working in your garden shed it would be overly ambitious to attempt to produce the next generation of computer chips. Intel or AMD are much better equipped for such a difficult and costly venture. (Software might be a different kettle of fish though: a lone genius programmer might realistically attempt to develop the next useful tool on the internet, for example.)

Another of my early gadgets was a device I called, 'Photoscope'. This was back in the days of chemical photography when colour was more realistic on transparencies than on prints, so we all took slides. There were two options for viewing slides; you could blackout the room and project them on to a screen for an audience to see, or you could pass round a battery powered viewer for individuals to look into. Photoscope was like a brief case and when you opened the lid it pulled up a translucent back-projection screen as from a roller blind. The idea was you would put slides in near to where the handle was attached, the beam would be reflected from the hinged side of the brief case to shine through the screen and a room full of people would be able to view them in daylight conditions just like watching a television. Little Photoscopes on the mantelpiece could have shown an illuminated picture of your grandchild for example. It would have been a boon to teachers, commercial travellers, photographers, etc. Nowadays we would use a laptop or the, recently launched, digital picture frame which allows your camera-phone to download to it by means of blue tooth wireless technology.

I was a young science teacher when I spent most of one summer vacation trying to make the innovative optics of Photoscope work (I'm not telling you what the innovation was – I might get it made one day!), without the benefit of an optical bench. I couldn't do it. That was when I realised I needed to **invent something simple and cheap enough to develop at home.** Even this advice comes with a caveat; I've just been refused a grant to develop a product I've already patented on the grounds that the level of technology is too low!

6. Low production costs.

Business is all about buying cheap and selling dear (and pocketing the difference). **You want as big a gap between the cost and the price as possible.** There are capital costs and running costs. In the case of Ideasun products the capital costs are for things like moulds, dies and cutters for manufacturing in plastic, and the running expenses are the cost of the plastic raw material and the charge for operating the machine.

At Ideasun we haven't gone into manufacturing ourselves, it's sub-contracted out. This is the modern way. Arkwright may have invented the factory (it was known as a 'mill' in his day because the early ones were powered by a waterwheel) to house his spinning frames, but that is because he was the first industrialist so he had to create this new type of building. Nowadays we recognise that there already are people set up in factories, with experience and skills in all varieties of machining and production. Don't try to 'reinvent the wheel' yourself, unless you want to become a machine minder, go to one of these existing experts. Better still go to at least two and get quotes; you will be amazed at the difference in their estimates. Don't be surprised though if the final amount you have to pay is more than the quote; things can change during tooling up and learning how to make your product. The world price of the raw material may have increased, especially if it is derived from oil or another commodity, and problems encountered during machining may have necessitated a change in the process.

Later, when you are a major international business, then you can look at 'vertical integration' and buy up your suppliers so that their profit becomes yours. Beware though, this is a difficult step and several quite big boys have tripped up over it. The Body Shop, whose expertise is in cosmetic product development and presentation, went into manufacturing for themselves and eventually had to sell off that arm of the business when they discovered that it would cost them less to buy the stock from a supplier *who had other customers so he could run his machines all day and all night.* Unfortunately, one of his other customers was L'Oreal who, spotting an opportunity, went on to make a bid for The Body Shop itself. In business, nothing is sacred.

Most of all don't waste time and money going ahead with manufacturing until you have established that there is a market for your product (see Trial Marketing). If you can do this as a 'cottage industry' it will really keep your investment down (see Prototype Production). Don't forget also that cost to you is not just how much you paid for the raw materials and the manufacturing. There are all sorts of 'invisible' costs including telephone, car, computer, accountant fees, patent agent fees, and, one that is likely to be the largest expense of all, *marketing* (more about this later). Get some quotes and do some sums before giving up the day job.

7. Direct sell suitability: Small size and low weight.

Do we want to share the fruits of our invention with lots of importers, wholesalers, distributors and retailers? I think not. So we have to consider the implications of 'fulfilment' - the jargon word for delivering the customers' purchases to their doors. Little, light objects that will go through a letter box are better than bulky items that have to be delivered by a carrier who invariably calls, requiring a signature, when the householder is out. Watch out though because when you start shipping overseas the hauliers calculate 'volumetric weight' which makes expanded polystyrene or any light product hideously expensive to send!

The ideal product would be easily packaged and posted from a direct sell, mail order website and/or toll-free telephone fulfilment service in order to eliminate all those middle men who want to hitch a ride on your product and share in the profit margin.

Once you get involved with a whole chain of distributors, wholesalers and retailers you are in a situation where you have to negotiate away large chunks of your income. You might feel, like me, that this is unfair since the product is your idea and the assorted merchants have done nothing so clever or deserving. You will especially be appalled at the margin some high street retailers and out of town superstores expect to command. Just to get on a shelf in their shops will cost you up to 72% of the selling price with another 15% disappearing to the VAT man. You are left with just 13%. In times gone by, I believe the yardstick was: manufacturer's price = cost to wholesaler + 100% = cost to retailer + 100% = cost to consumer. So the manufacturer got a, slightly more realistic, 25% share of the final return. Those were the days and unless or until they come back I advocate cutting out the middle men whenever possible. Fortunately, today we have the internet – *get yourself a good website*. Ecommerce is growing at an incredible rate; an estimated £8.2 billion worth of goods were bought in the UK from websites last year, up 28.9% on 2004 while high street sales only made an extra 1.5%.

8. High perceived value.

You need to sell millions of packets of crisps (potato chips, if you are American) to make significant money because each packet has such a low price – just a few pennies. And that's despite the fact that they have an enormous profit margin. Do you realise that you can buy a sack of potatoes at retail for about £4 and this would make over 800 packets of crisps that you could sell for 50p each? Around 10,000% mark-up! (Ok, there is the cost of cooking oil, salt, fuel, packaging, machinery, premises, staff, transport and retailers' cut to take into account, but it's a great margin isn't it?) And crisp manufacturers can buy lorry loads of potatoes much cheaper than retail sacks, direct from the farmer. Will you ever buy crisps again? Go home and slice just one potato very thinly and toss it in the fryer for heaven's sake! It's probably eight packets of crisps worth. They even taste better! Learn how to slice potatoes thinly or buy a food processor to do it. If you want to throw your money away there's probably a drain in the road outside you could put it down, but please consider a genuine charity first.

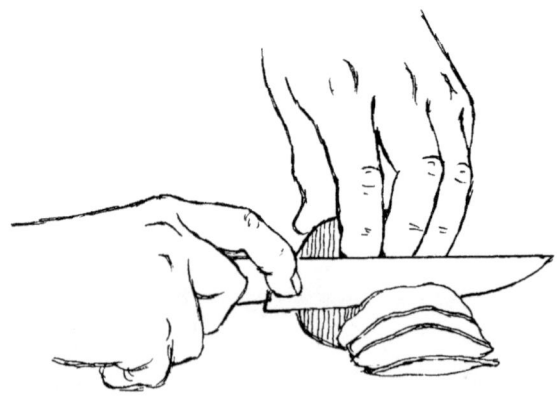

> **Home-made crisps**
>
> Heat some vegetable oil to 190 degrees Celsius. Wash some medium to large potatoes. Do NOT peel them! Cut them in half along the longer axis so that you can put them down, on a cutting board, on the flat side you have just cut. Holding the potato down with one hand, slice it with a sharp knife. Use the index finger of the hand holding the potato to guide the knife into making very thin slices. Keep that finger upright so the knife slides down the nail as you steadily move it back each time you cut. Fry the slices, stirring occasionally until golden brown. Remove and drain. Add sea salt and ground black pepper, rosemary or curry powder. Eat.

If sold direct to the consumer from a website or free-phone number, a product needs to have a price which can be put on a credit card without too much pain. However, once a product is priced at £300 (or thereabouts) most people are reluctant to buy it unseen and untested unless the vending company has an unimpeachable reputation, so maybe this is the ceiling.

To be worth trading, especially if dispatching by post or carrier, there is no point unless the price is at least £1.99 plus a shipping charge. Of course you can bundle several cheaper items together to reach the minimum threshold and sell, for example, boxes of 10 packets of crisps.

9. Repeat sales e.g. subscriptions, refills, resales, accessories

This may sound like a tall order but wouldn't it be wonderful if, having sold your product, the purchaser then had to pay you a monthly subscription to continue to use it? Or, perhaps, every few weeks some consumable item ran out and had to be bought from you? Maybe the product could become old-fashioned or outdated and need changing for the latest model. The point is **subscriptions or refill/replacement purchases make an excellent long term reliable income.**

Best to be aware, though, that you can take this too far; back in the nineteen seventies we had something called 'planned obsolescence'. You bought a car knowing it would turn into a pile of rust before seven years were up. Some manufacturers deliberately made things with a short life to make you buy a replacement. Many products were designed to last just a bit longer than the guarantee. Those manufacturers got the reputation they deserved and few resales. Many have gone bust.

That period of the twentieth century was a good time to do business in the West. The market was expanding; few people in war-torn Europe had possessions like cars, freezers or colour TVs. Everyone wanted them and we were all working hard and saving or borrowing to be able to buy them. In a boom time like that selling is a little easier. Quality could be poorer. Now most people in the developed world have practically everything they want so, in order to do business, all manufacturers actually have to try to make their products better than the competition or offer something completely new like an iPod. Today, of course, there are expanding markets in the developing countries... It's going to be really tough trading when they have caught up...

10. Finally, consider timeliness.

Your invention must be something that people want *now*. There's no point hoping that, in the future, buyers will come to realise its value, or in providing an improvement to yesterday's technology. In the first case you might not live long enough for consumers to agree with your concept of potential worth and in the second, even if you do come up with a better valve (tube) amplifier for electric guitars, for example, you will be unable to prevent its decline as digital signal processing continues to take over. It's just as bad being ahead of your time as it is being out of date.

Thinking about that summer vacation spent failing to develop my Photoscope prototype for displaying slides has reminded me that an invention may have its time. Technology has moved on; there is hardly any market for chemical photography related products anymore. The UK high street trader, Dixons (now Curry's digital), has only sold *digital* cameras for the last few years and has recently discontinued videotape machines in favour of DVD players/recorders. If George Stephenson was to come back to life he would be astounded to see there are no steam locomotives used in England any more, except in the nostalgia industry. Consider this; now that most food and drink cans have ring pulls, you can stop devising new can openers. Don't invest in Compact Disc technology now that everyone is going over to MP3s stored on hard drives or in flash memory. I expect you get the message!

But, if you have some hot idea, don't waste a moment or you might be left behind by events! Race it to the market. Don't leave opportunities for others to invent the same gadget or for the market to change so that it becomes unnecessary. Learn from the lesson of 'Baby Dream'; two guys who appeared on the BBC2 venture capital programme 'Dragon's Den' pitching for investment in their baby buggy rocking device. They failed to get investment because they refused to part with enough of their business, but they came back the next year with another bid. The trouble was, in the meantime, someone else had brought a buggy rocker to the market so the business 'angels' were no longer interested.

> When asked by a young journalist what can most easily steer a government off course, **former British Prime Minister, Harold Macmillan** said, *"Events, dear boy, events!"*

Almost unbelievably, every product has a lifetime or, if not a lifetime, then a heyday – even whisky and cauliflowers can go in and out of fashion and that's where marketing comes in; see later...

Ideal Product Warning:

A product that meets all ten criteria still gives no guarantee of success; even the best product is prey to the beast of marketing! That's the most difficult and expensive part of any 'concept to consumer' project.

Now let's take a look at some existing products:
Let's see how well they fit the ten criteria.

Example 1: Scent

The appeal of scent is mainly to women. It hasn't really caught on so much with children or the male half of the adult population, except for pubescent boys who get paranoid about armpit odour. That's a huge, almost untapped market, and this is despite several attempts by the manufacturers to sell to men. A bloke can hardly go into an upmarket male fashion store these days without being accosted by someone (usually a hot babe) wanting him to try their fragrance for men. It amuses me because the only word they can use to describe the product's odour is 'fresh'. Cosmetics for women can smell of fruit or flowers, spices and musk but men don't want to smell of those things so that just leaves 'fresh'!

You can put a big tick in the boxes for minimum investment and low production costs though. Dame Anita Roddick started her worldwide Body Shop company by manufacturing cosmetic products on the kitchen table of her home in Brighton. Scent costs very little to make but you might have to spend a fortune on packaging and advertising to get it to sell. It's a mature market with established brands that have an enviable reputation for that indefinable quality: glamour. Just try giving your girlfriend some unknown label perfume and you will find out.

> **Scent**
> Meet a want ✔
> Not like existing ✘
> Not too far out ✔
> Max appeal ✘ (*not men*)
> Low dev. costs ✔
> Low prod. costs ✔
> Small size, low wt. ✔
> High value ✔
> Repeat sales ✔
> Timeliness ✔

Bringing a new brand name to that market, like Anita did, was an incredible achievement. She targeted the zeitgeist of the 1970s and it worked. Her insight told her that people wanted cosmetics that were made from more natural ingredients which had not been tested on animals (recently!). Her packaging was basic and refillable, her marketing budget nil. She tapped into the trendy young community wandering about the streets and lanes of Brighton. The fragrance emanating from her shop drew them in. It resonated with the spirit of the age which was all about hippies, self sufficiency and environmental friendliness. The chain of 1,900 Body Shop stores in fifty countries worldwide grew from that single shop squeezed in between two funeral parlours back in 1976. That moment has gone. The pendulum has swung. You probably couldn't launch the Body Shop in today's profligate, sports utility vehicle world. Eco-friendly may not count for so much anymore, but the population is even more self-indulgent than we were in the seventies, so the company continues to do good trade now it is established despite being owned by L'Oreal who aren't exactly famous for making kindness to the environment a sales point.

Since I wrote that George W Bush has suddenly gone a tiny bit 'Green' so perhaps there is hope after all!

The Author

Scent is one of those products that we want to try before we buy so the high street perfumery departments or airport shops are still the best outlets. Pricing it is one of those 'think of a number' exercises. If you can associate a scent with a glamorous celebrity and put it in a fancy bottle you can charge what you like. That's why new celebrities launch a scent as soon as they have a name to trade on.

In fact, for perfume, the rule is the costlier the better: it gives it the cachet of being exclusive to rich folk; that desirable club you don't belong to, but at least you might be able to afford to smell like the club-members! There is one fragrance in a genuine ruby jewelled bottle that sells at £115,000 but if it was cut-price you'd be suspicious that it was a fake. It's like diamonds: no-one wants to be given a cheap one! Repeat sales are a possibility but only at longish intervals.

Example 2: Mobile Phones

There are more cell phones in the UK than there are people! Sixty million! I can't think of many products which have better market penetration. They can be sold direct to the end user and delivered by post. The price falls into the credit card purchase range and the customer has to enter into a subscription agreement requiring him/her to pay monthly for ever and a day.

Mobile phones	
Meet a want ✔	
Not like existing ✘	
Not too far out ✔	
Max appeal ✔	
Low dev. costs ✘	
Low prod. costs ✘	
Small size, low weight ✔	
High value ✔	
Repeat sales ✔	*subscription*
Timeliness ✔	

Ok, he's not paying the manufacturer directly for the service (they have an arrangement with service providers), but what an amazing product! There are even accessories like fashion covers and headsets which can form a secondary market. The phones are constantly being improved so you have to buy again in a year or so to get the latest features. However, they are not much of a prospect for the individual inventor – development and manufacturing costs must be vast – we will have to leave Nokia, Motorola, Samsung and Sony Ericsson to compete for that business. Even those guys can misjudge things; they've produced some weird cell phones (e.g. the swivelling Motorola V70) that were too far out to sell in large numbers.

A friend of mine did come up with a nice idea for a combination mobile phone and removable front to a car radio/CD system. The patent cost a small fortune and he has yet to get interest from the big companies. If anyone reading this wants to run with his idea I'll be happy to put them in touch with him.

Example 3: Computer Printers

Printers have a maximum market of about one per home or office, so not as many as mobile phones, but the constant need for ink ensures a continuous income for the vendor. They can't be posted through letterboxes so are mostly sold from stores or delivered by a carrier. Development and manufacturing costs must be huge so there's not much here for the inventor. The interesting thing about them is the way manufacturers have massaged the price of the printer downwards, and the ink upwards. It's almost cheaper to buy a new printer when the ink runs out. They are obviously making nothing on the printer in order to get you to sign up to their system then recouping the money by inflating the price of the cartridges. Counterfeit refills may be a threat to this cosy expectation.

> **Computer Printers**
> Meet a want ✘ (*a function*)
> Not like existing ✘
> Not too far out ✔
> Max appeal ✘ (*not personal*)
> Low dev. costs ✘
> Low prod. costs ✘
> Small size, low weight ✘
> High value ✔
> Repeat sales ✔ (*ink refills*)
> Timeliness ✔

Example 4: Golf Tees

Dr. Venanzio Cardarelli, a dentist in Plymouth Massachusetts, has invented a tooth shaped golf tee. His tee is shaped like the crown of a molar so it makes less contact with the golf ball. He's not the only one; there are dozens of versions of golf tees including one with three roots, again like a molar. If they do confer an advantage it's going to be so minimal that there will be little incentive to purchase the new version rather than the established product. Even so they are only going to appeal to the golfer; there may be millions of golfers

> **Golf tees**
> Meet a want ✘ (*a function*)
> Not like existing ✘
> Not too far out ✔
> Max appeal ✘ (*golfers only*)
> Min dev. costs ✔
> Low prod. costs ✔
> Small size, low weight ✔
> High value ✘
> Repeat sales ✔
> Timeliness ✔

but they can only be a small percentage of the *total* world population. I hope these guys can sell billions of their innovative tees because, the unit price is so low, that is what they're going to need to do just to recoup the cost of the patent!

Example 5: Greetings cards

When I was a boy you could buy Christmas cards and birthday cards. There was a small display in the newsagents or confectioners. That was it. Now there are cards for mother's day, father's day, valentine's day, engagement, wedding, anniversary, get well soon, good luck in your exams/driving test,

Greetings cards
Meet a want ✔
Not like existing ✔
Not too far out ✔
Max appeal
Min dev. costs ✔
Low prod. costs ✔
Small size, low wt. ✔
High value ✔
Repeat sales ✔
Timeliness ✔

congratulations on your new baby/passing your exam/driving test/winning the match, going to school/university, your new home, pet's birthday, sorry you are leaving, retirement, condolences, in sympathy, etc. etc. Most high streets have a large shop dedicated to retailing just cards and a lot of superstores have an entire aisle. They are selling pieces of card with printing on, charging exorbitant prices and the public can't stop buying! Hallmark has diversified into an enormous shopping centre in Kansas City and has a very popular TV channel. What an amazing industry! Cleverly, they are not simply selling printed card; they are selling *thoughts*. Customers will go along with their huge mark up because 'it's the thought that counts'. Costs are tiny; the original is generated on a computer and emailed to the cheapest printer anywhere on Earth. A 'print farmer' will find you the best price if you can't be bothered to search yourself. Printing is a long established and very competitive industry so there are few problems of manufacture and high quality is easy to achieve. The products are not heavy or bulky which means storage and transport are a low cost item and there are vast opportunities for growth in the developing markets around the world. The only difficulty is in getting started; it may take you years to become established in such a hotly competed market.

Summing up, the aim is to invent a product with a decent profit margin that the maximum number of people want to possess now and can afford to buy. My advice to you, as you go about your daily life, is to take a look at the items currently available on the market and try to analyse why they are successful. See how many of the ten criteria they meet; I can claim that our Easyriter pen gets ticks in every box. *You* have to invent something that hits enough of the right buttons to make a fortune too.

> *The best sort of business is one where you can sell a product at a high price with a good margin, and in enormous quantities.*
>
> **James Dyson**

> *Art is pleasure, Invention is treasure...*
>
> **Trevor Bayliss**

PROTECTING YOUR INVENTION

As with most things involving lawyers, this is a quagmire. The only safe place to keep your invention is in your head, as a secret idea!

Once you want to turn it into a product it has to be presented in a written or pictorial form and must meet certain criteria. This may sound cynical but, naturally, the legislators have come up with different laws in different countries.

The next most safe situation to be in with your product is to be manufacturing it at so low a cost that nobody thinks it's worth setting up to compete with you. Failing that you need to get some intellectual protection and that means a patent, a design registration or copyright.

In the UK, to be granted a patent, your idea has to be innovative and it must not be 'in the public domain'. The thinking is that, if other people already know about it, how do we know it's yours? So, don't go telling anyone without first getting them to sign a non-disclosure agreement (get one from www.ideasun.com).

It's not like that in the USA. You can apply for a patent up to a year after you have published the concept and they are 'liberalising' the system further. This is despite the fact that the US Patent Office is drowning in filings and getting known as an 'easy grader' through awarding patents too leniently. The outcome is a stifling of innovation and a bonanza for litigation; US patent lawsuits have increased 58% in the last ten years. The system obviously needs reform although I don't suppose the lawyers will agree. I wrote the above in 2005 and revisiting this paragraph at the editing stage shows me I need to update. The US patent office has recently become much stricter. In fact the pendulum has swung the other way; they are rejecting applications with excessive vigour! Now the lawsuits will be in the form of appeals against rejection. Lawyers always win!

Many inventors question why they need to pay renewal fees every year once their patent has been granted. What work does this pay for? Song composers and book authors have no such charges for their intellectual property and their ownership is for life plus fifty years (seventy years in some countries). Why can't we have a level playing field?

In all countries your application is checked against the 'prior art' to establish that what you are proposing is genuinely innovative. You can do some of this yourself; log on to www.esp@cenet and search for keywords relevant to your idea. The patent examiners can reject your proposal if 'a man skilled in the art' (who's he?) might consider it obviously derived from an existing patent. Essentially, you need a patent agent, then the clock will start ticking and the bills will start coming in before you've earned a single penny from selling your great idea. It's a countdown situation because patents expire twenty years after being granted. (Compare that with copyright!)

The best parts of the world to get protection in are those with the biggest market i.e. a large population of rich people. This currently means Europe and the US, but keep an eye on India (*1.6m Indian households now spend $9000 a year on luxury goods* Time April 10th 2006), the old Soviet republics and other countries where economic development may be producing more disposable income for their already huge populations. Lots of other countries have one of the two factors but not both. For example, New Zealand is rich (GDP per capita $20,000) but small (population 4 million), while Brazil has a large population (200 million) of mostly poor people scratching a living on $2 a day, apart from a few jetsetters in the cities of Rio de Janeiro and Sao Paulo. The hope is that, if you have control of the best markets, a rival might think twice about whether to bother to tool up and compete for the poorer territories.

It's good business for patent agents and examiners, I suppose, but for the would-be inventor these national or regional differences mean that you have to make applications in several countries, or groups of countries like the European Union. Some territories, particularly China, have only recently signed up to the notion of protecting intellectual property and have negligible enforcement, so there's a third of the world's population who can steal your idea with impunity. While they are manufacturing your order they are shipping your product out of the back door to their own distributors. If you object they simply relocate – you try to find them in the vastness of China!

In order to give the inventor the greatest opportunity to develop his invention, its description in a patent needs to be as broad as possible; you want to prevent a competitor from making something slightly different that does the job just as well. This in itself can inspire inventiveness; the sun and planet gear system for turning reciprocating into rotary motion was created to avoid infringing an existing patent on the crank.

A few patents are prevented from publication by the government on the basis of state security. We don't want the enemy to have our ideas for weapons or ciphers, do we! Stupidly though, the United Kingdom Government is not flawless. They failed to see the significance of Frank Whittle's jet engine, offered him no financial support (in typical British fashion) and allowed the patent details to be published. Starved of cash, he took longer to develop his invention than the German, Dr Hans von Ohain, who, although his slightly different patent was registered a whole six years later, lived in a country where its value was realised and aircraft builder Ernst Heinkel poured money into the project. Fortunately for The Allies it was still not ready in time to make a significant contribution to World War Two. (I would like to think that, in today's global village, the concept of warring nations, especially offensive invasions by democracies, is history but George W Bush seems to have other ideas. I can't wait for nationalism/religious fundamentalism to die and global sanity to break out. Whatever happened to the Love and Peace mood of the Sixties?)

Some inventions are very difficult, which means expensive or impossible, to patent. Take Trevor Bayliss' Clockwork Radio; it's just a hybrid (sorry Trevor). Clockwork and radios both already existed. It's not much more innovative than a car radio or a radio cassette. Another example of this sort of claim for novelty was the first patent for attaching an eraser to a pencil which was issued in 1858 to a man from Philadelphia named Hyman Lipman. His patent was later held to be invalid because it was merely the combination of two things, without a new use. Trevor's idea does have a new use because it was the first to claim a radio *powered by* clockwork but my patent agent has told me that the examiners decided it's not patentable, although it is a very successful product. The upshot is, I wouldn't spend much money trying to protect the 'umbrella radio' or the 'binocular radio' if I was you.

Surprisingly, patents do not have to be *granted* in order to be effective. Just having *applied* for a patent can scare off or, at least delay, some competitors from entering the fray so you should make sure your products have 'Pat Applied For' or 'Pat Pending' prominently written on them.

Piracy

If you do get to be the proud owner of a patent, don't imagine that you can rest on your laurels. If your invention is a commercial success it will certainly be copied whether protected or not. In fact, if your idea is *not* challenged by copyists it probably isn't an attractive economic proposition.

James Dyson CBE, inventor of the cyclone vacuum cleaner, had a battle with Hoover who rushed a strikingly similar, bagless, cyclonic model into production. In 1999, he was forced into court to protect his invention. It took 18 months for James to finally win a victory against Hoover for patent infringement. Ironically, they were one of the many companies who scoffed and showed him the door when he offered to license to them years before. They already had a good business, why should they waste money tooling up for a new product which could make their existing range redundant? Especially when the likelihood was that Dyson would never be able to afford to manufacture himself, without the support of a company like them, and therefore was no threat to their sales. Vacuum cleaning efficiency didn't enter into it! But they didn't reckon on his bulldog tenacity.

On a much more modest level I have already had to pay lawyers to combat a pirate who copied our Wordwiza reading tool. What happened was this: I sold more than three hundred Wordwiza at a London Show and then a few days later I had a phone call. This guy said his partner had bought a set of Wordwiza from me and they thought it was a good product so he wanted to know if we could supply at a wholesale price for him to retail off his website. We had several telephone and email communications discussing possible terms of business and then suddenly I heard no more. When I emailed him asking whether he had got my last email he responded saying, yes he had, but their circumstances had changed. Thinking maybe he'd gone back to the day job, I left it at that.

Six months later I checked out his website and to my amazement there it was: a home page advert for his counterfeit reading tool which he was claiming was 'new and exclusive'! Same four colours, same price, very similar flash animation, everything! I sent him an email letting him know that he was infringing my patent and that, unless he agreed to my licensing terms, I would have to take legal action. I attempted to get the Trading Standards people to deal with him but it turns out that the legislation empowers them to deal with breaches of copyright or trademarks, not patents! How stupid is that? It's another way in which inventors are not treated on a par with authors or songwriters.

So I had to get my lawyer to scare him off. The lawyer bought one of his fake reading tools from him to have as a piece of evidence. It was so fragile it soon got broken into two halves and I've got the pieces in our 'black museum' now. Yesterday I was gratified to hear on the internet that his dotcom company has ceased trading; one victim of the business jungle who gets no sympathy from me! Since writing the foregoing, I've just met, at the BETT 2006 Show, some education advisors who recommended his product and others who complained that it soon got scratched. When I told them the story they apologised for having supported this cheat and changed their recommendation to our product. They were especially impressed when I showed them a real Wordwiza can be bent back double or almost tied in a knot with no ill effects. (Of course, everything has a downside. We will get no replacements market for our indestructible product.)

Imitation may be the sincerest form of flattery but the only ones sure to benefit from protecting the rights to a product are the lawyers. Defending a patent can come to a trial of strength with the prize of possession going to the side which can afford the best lawyers and pay for the longest court case. If you do get involved with litigation you will need access to vast funds unless the contest is in the US where lawyers work on contingency – no win, no fee. In the UK the strength of a patent is directly proportional to the depth of your pockets.

> **Have a fighting fund**
>
> *We were lucky, in the early stages, that we didn't have any legal cases to contest. Looking back, we should have made sure we had the funds to police our Intellectual Property rights. The value of IP protection is diminished if you can't afford to take anyone to court.*
>
> Nicki Owen, Trainique Ltd

Patent disputes

As an example of the sort of conflict patents can get you into let's examine the case of the blue LED. Light-emitting diodes were invented in the 1960s. Red and green ones quickly found a market but the blue LED was the Holy Grail; it would open the door to semiconductor white light, full-colour displays and high capacity DVDs written by a blue laser. Finally, the first blue LED was launched by a Japanese company in 1993. Then in 1999, an employee left his job and claimed he had invented the blue LED single-handedly. He moved to California and sued his old employers saying they'd only paid him $200 for his great work. In January 2004 a court calculated that, by 2010 when the patent runs out, the company would have made $1.2 billion from the invention so they awarded the claimant $190 million. The firm appealed and in January 2005 another court ruled that the company was responsible for 95% of the invention and said the claimant should accept $7.6 million in compensation. The dispute rumbled on, giving regular employment to lawyers and I've just heard (Jan 2007) that he has settled. That case has changed the policy of companies in Japan.

Copyright etc

There are other means of protection though. Somebody must have the rights under registered trademark law to 'Freeplay', the name of the company that makes the clockwork radio. Written property, including software, can be protected by copyright and, of course, you can own your website name. The shape of a product or artwork of a logo can be protected by registering the design which can be done up to one year after public disclosure. For impecunious novice inventors a good organisation to belong to is ACID. Anti Copying in Design is an action group committed to fighting copyright theft see: www.acid.uk.com

Don't forget that you can sell a patent or registered design; I know that, as an inventor, you will be loath to do this but, if your personal circumstances get desperate it's nice to have in reserve.

MAKING A PROTOTYPE

Following my experience with 'Photoscope', when I realised that I needed to be able to do things 'in-house' as far as possible if I was to get anywhere other than to be at the mercy of a rich company, I tried to focus my attention on gadgets that could be produced, to a standard suitable for trial marketing, on the dining table or garage workbench. I was lucky in that I came in on the end of the period when a lower standard of finish was considered acceptable by the buying public.

Ten years ago my reading tool was made at home, by my late father and me, using strips guillotined from a sheet of plastic and an electric bar fire to soften them for bending. (You can see these on our website www.ideasun.com) When a strip had been hand shaped into a reading tool, it had tinted labels stuck on it, was latexed to a white card which had been printed in black ink and then sealed in a plastic bag by sellotape. We included an A6 instruction booklet, made from a single sheet of A4, which had been laser printed on both sides, guillotined into two A5s and stapled by hand using a saddle stapler. This was not merely a prototype; it was good enough for trial marketing. Thousands of those were sold at a considerable margin before we tooled up to have them injection moulded. It was very downmarket and simple, but I still get requests for this version which I supply from old stock. I ought to thank the boss of a school supplies company in Partridge Green who advised and encouraged me, and gave me the original sheet of plastic that we made them from.

Hundreds of our teaching clocks, called 'Cloxi', were made originally by the same team, on the same dining table. I sourced a plastic box and lid (like a Petri dish), originally made for another product, and we adapted them. The clocks went through a revision during this prototype stage when we improved the hands and knobs and upgraded to two colour printing. I've only ever had one complaint about it and that woman thought the concept was good but said her five year old child expected the quality to be 'a bit better' so I gave her a refund. She has asked to be notified when we have new supplies. We had sold out at the time so we welcomed her returned clock which gave us a stock of one and I soon resold it. Gratifyingly, Cloxi has been used to teach Downs Syndrome children how to tell the time. We thought "Great! We'll get it on the shelves of the Early Learning Centre." No dice. They wrote back: "We've got our own in-house design team who provide us with a constant supply of great new products."

Both the clock and the reading tool were pre-production versions that could be trial marketed to establish the viability of the products but, today they probably wouldn't cut the mustard and you will have to risk thousands of pounds tooling up to produce a high street shop level of quality just to discover whether folks want your gadget or not.

Our next product, the Easyriter pencil, was a more conventional prototype in that it was merely a mock up, not something that could be trial marketed. Although we could see that an improved pencil, and subsequently a pen, could meet all the ideal product criteria, it didn't take a genius to work out that we couldn't make quantities of pencils or pens to a satisfactory standard on the dining table.

I'd been thinking about pencils for children since helping my first wife (very sadly, Pam died of cancer in 1998) to tidy up her infant classroom. The pencils she was using with the children were larger than conventional ones: 11mm instead of the normal 8mm in diameter. In a five-year-old's hand they were almost proportional to a broomstick in the hand of an adult! Unsurprisingly, the children were *holding* them like broomsticks. We conducted a survey by putting out a selection of pencils for the children to choose from. Out of a range from fat to thin, round, hexagonal and triangular, the one they chose was the skinny pencil that slides down the spine of a pocket diary. When we asked them why they had selected that one, they said, *"Because it's small like me!"* It seemed like irrefutable logic.

I worked out that a pencil with a small core, and fins on it to fit the shape you make between the fingers and thumb when you hold a pencil correctly, might train children to adopt the best grip. Life took different turns and it was years later (2002) that I applied for a patent and made a prototype using a propelling pencil lead and the sort of modelling plastic that you set hard by baking it in an oven. I fashioned it with three concave facets but, when I took it out of the oven I discovered that it had softened before setting and gravity had caused the underside to flatten against the baking sheet. Richard, my business partner, pointed out that this actually fitted the hole between the fingers even better. This is an example of serendipity. There ought to be three names on the patent application: mine, Richard's and the 'father' of gravity, Sir Isaac Newton.

I made several more pencil mock-ups in wood using a router, in order to check on different dimensions and profiles, then got a graphic designer to do a scale drawing which I could take to a manufacturer. This story will be continued in the section on manufacturing.

You shouldn't have so much trouble because, thanks to technology, prototyping is much easier these days. Many products are suitable for making on a computer controlled machine. You just feed in the design file and watch it happen. All it requires is money! Put CAM into Wikipedia search to find out more.

Professional Design

Don't imagine that you can miss out this stage. Once upon a time you easily could, but now that machine tools are mostly computer controlled you need a CAD file to run them. And the service that the design houses offer is really good. They not only bring their viewpoint and expertise to bear on your product but they can provide you with useful contacts like manufacturers or agents for overseas production. For a modest fee they will produce coloured and detailed printouts of your invention, or files that will rotate on your laptop and impress the bank manager.

Some manufacturers will claim to be able to do the computer files necessary to operate their machines. In this case, *you* will be the designer issuing instructions or okaying proposals which will be committed to a machine tool. Don't you think that it might be a false economy to miss out on in the opportunity to collaborate with someone who has specialist knowledge of design?

I'm a firm believer in the proposition that two heads are better than one so I welcome the input of designers who might investigate competing products and make suggestions about how to aim your goods at the market, taking into consideration such things as brand identity and 'coolness'. It's much better to get your product and packaging finalised at this stage than to have to scrap stock and retool the manufacturing when the, badly designed, first batch breaks or simply doesn't sell because it didn't appeal to your target age group.

Stereo lithography

I couldn't write a section on making prototypes without mentioning stereo lithography. This is a wonderful new technology that allows you to 'print' in three dimensions. It can build up, layer by seamless layer, a physical model from a CAD file. There are several versions that allow you to use different plastic materials to produce a sculpture of your product. It can be very useful to be able to handle an embodiment of your design and get a feel for the balance etc. It is also an excellent presentation aide when seeking support from financiers. We had them done for the Wordwiza and the Easyriter pen. It's so close to the final product and yet you can still change it, if necessary, before committing to tooling up for manufacture. Soon people will be making small runs of high value finished products in this way instead of injection moulding. The only problem is a stereo lithography machine costs in excess of £25,000 at the moment.

There is also another technology that uses a computer controlled router to produce a physical embodiment of CAD files in plastic, wood or metal. I hope your prototype can be made in one of these ways.

Naming Your Product

A new invention can have a newly invented name. What fun you can have putting words together. How did they come up with 'Marmite' for the processed yeast sediment that used to be thrown away at the end of the brewing process? The company aren't very sure but say it may have been named after the small earthenware casserole dish it was originally supplied in: 'une marmite' (pronounced 'marmeet') in French. What about Bovril? That name has two parts, *bos* which is Latin for 'ox' and *vril* which was 'an electric fluid that cured diseases and established equilibrium of natural powers' according to a popular 19th century novel featuring a subterranean humanoid race. (Bulwer-Lytton 'The Coming Race') Anyone interested in the origins of product names can have hours of pleasure checking them out on the internet.

Even here there are pitfalls though. Toyota can't market the MR2 in France because its designation is pronounced 'emm air deux' and sounds like 'merde' – French for 'shit'! Our first attempt at a product name was too clumsy by far. We called the reading tool a 'Libli'. It was an acronym for 'Little in Big Literacy Improver' because it enables you to find little words inside big words and I thought it would fit into the child's vocabulary along with mummy, teddy, cuddly, etc. Apart from being hard to read and say, especially if you are Chinese ('ribri'!), it's not all that memorable. We soon changed it, at Richard's suggestion, to the much simpler and more logical 'Wordwiza'.

You go ahead and enjoy creating a name for your product but don't forget to road test it on your intended market – it should be meaningful and memorable to your potential customers.

Your Company Image

More fun can be had deciding on a brand image; a company name, logo and strapline. Obviously you want something memorable and possibly relevant to your product. Some people try to get to be first in the yellow pages by calling themselves 'A1 traders' in the hope that you will stumble on their name and go no further. Taking that strategy to extremes, 'Aardvark Associates' would be the ultimate!

You have to decide whether to give your company an explanatory name: 'Fabulous Bakin' Boys' gives a pretty good idea of what a firm does but calling his company 'Virgin' never did Sir Richard

Branson any harm despite the supplying of virgins not being included amongst his many products and services! When it was started, back in the sixties, 'virgin' was a risqué word and therefore unforgettable. So there are no rules here – just have fun.

Bear in mind though that creating a brand is an even more difficult and *long term* task than inventing a commercial product. Don't make the mistake that the recent team of contestants in BBC Two's 'The Apprentice' did when they were given the task of selling cups of coffee on the streets of London. They named themselves 'Eclipse' and shook chocolate powder through a stencil so that it left a print of their logo on the froth of every cup which they decided would then be a 'coffee experience'. Why did they waste time and effort trying to establish a brand during the only day they would ever be in operation? Probably because their professional experience was in marketing. This was a completely redundant skill for the task in hand; they needed to sell, not to market but they didn't know the difference!

A company I used to deal with made another silly mistake. They called their website 'imaginationict'. We were all saying it like 'imagination' with 'ict' stuck on the end. It was only when I heard the owner say the name that I realised he thought it should be pronounced 'imagination eye cee tee'! He was selling a service under one name and his customers were buying it under another! He'd deliberately confused everyone!

Most well known brands were established early in the history of commerce; Sainsbury's was started in 1869. The majority of recently formed brands don't provide manufactured products but services associated with the internet like Amazon, Ebay, Friends Reunited, Google, Myspace, Wikipedia, Yahoo and Youtube, or online insurance, travel and loans. How many famous *manufacturing* brands that were created in the last forty years can you name? Apple, Hewlett Packard, Dell, Bodyshop: that's only four and I'm already running out of steam. Defunct brands are much more common, just think of all those extinct British car manufacturers: Alvis, Armstrong Siddeley, Austin, Hillman, Humber, Jowett, Lanchester, Morris, Riley, Rover, Singer, Standard, Sunbeam, Triumph, Wolseley, need I go on?

For a brand it's a case of the bigger the better so companies keep merging, using the most famous name and allowing the others to die. Small companies never get known beyond their immediate customers.

So the sad fact is, it's much easier to sell your invention to an existing company which is already well above the parapet in terms of public consciousness and which already has established trading connections with both suppliers and purchasers.

MARKET RESEARCH AND TRIAL MARKETING YOUR PRODUCT

Market research

Do you know how much market research Apple does when it introduces a new product like an iPod or iPhone? Zero! Absolutely none! They have a different strategy: total secrecy followed by a sensational press launch. If you have creative talent like Steve Jobs and his English designer, Jonathan Ive, why would you consult a focus group of the numpty proletariat?

Let's face it market researchers couldn't even correctly forecast the 1991 UK General Election result two days before polling day and all they had to do was ask enough people in enough places how they were going to vote. If you want to, you can spend a fortune discovering that 33.33% of mothers with children under 6 either 'quite like' or 'really like' Sainsbury's brand image orange colour and 17.66 % are undecided while another 14.77% would prefer it to be a slightly different shade – maybe candy apple glory red...

Market research is a great way of amassing statistics which look like a lot of work has been done and can therefore justify a huge fee but, unfortunately, the results have to be interpreted and, quite often, they signify nothing. Maybe it is useful if you are trying to determine which is the best available Cumberland sausage but be aware that Pepsi consistently outperforms Coke in blind tastings and yet Coke still has by far the biggest share of the market so what do you make of that? In my view, market research is for people who like questions more than they like answers! So it's optional and it's expensive.

If you don't understand your market, perhaps you shouldn't be attempting to make products for it. Market research is no substitute for a good gut instinct. Trial marketing on the other hand is a different kettle of fish...

Trial Marketing

I've put this section before manufacturing because, if you can possibly do your trial marketing with a pre-production version of your product, you will learn valuable lessons which may either stop you wasting money on going ahead with the manufacturing, or enable you to make your product even more desirable before committing to tooling which is expensive to alter. Certain products, like the Easyriter pencil, cannot be tested in this way because they need to be properly manufactured before you can put them on sale; although I did put the prototype into the hands of teachers at the Education Shows to get a reaction (see Manufacturing). For those types of product you will have to reread this section after you have achieved the manufacturing.

Obviously you had an end-user in mind when you invented your gadget but you need to check whether your instinct matches reality. You may think you have invented the ultimate in folding golf bags but one question is, 'Will golfers consider it sufficiently better than their existing bag to want to change over to it?' Another question is, 'How much will they pay?' There is only one way to answer these questions and that's by trying to sell it. It will be fun! Let the showman inside you get out for a while and discover how your potential customers like your product. See their reaction when you try out a price. Discussing how it may be used with them could even give you ideas for further improving it.

Trial marketing completely changed the direction of our first educational product. My infant teacher first wife and I had, as we thought, 'increased the value' of the original version of the reading tool by producing a pack of items for parents to help their children learn to read. We'd been worried that we might not be able to charge much for a tool which, whatever it could do, was in reality just a bent piece of plastic that I'd made at home. The 'Reading Key' pack had a book called 'Teacher's Secrets', a convex triangular pencil, paper with special guidelines, a poster of activities and the reading tool. Try as we might, we couldn't sell it and we had to learn the lesson that modern parents don't have the time to, or don't want to, help to teach their children to read; they believe that's what schools and teachers are for. In the UK the state pays for and, therefore, has accepted responsibility for education so many parents think they shouldn't need to get personally involved. In fact, they feel they can complain to the government when they don't get their 'rights' provided satisfactorily. The best clue about our product came when someone pointed to the reading tool in the corner of the pack and said, *"Could I buy that on its own?"* So we threw away all the other stuff and packaged the tools individually. Folks surprised us by willingly paying £3.99 for my home-made bent plastic!

Will they want to buy your product?

Trade shows or street markets are good locations to get your products in front of an appropriate section of the public to see whether they are sellable and what price they can command. There are shows for all different interest groups throughout the year and all around the country. Some charge a lot for space, some nothing at all. Beware though, there are pitfalls at every stage of the invention business. You may remember, earlier in this book, I told you about a failed Education Show that few visitors attended? It spelled doom for one trader. She had invested all the remains of her budget in exhibiting at that show and got nothing back. That was the end of her little business. Another exhibition I heard about, but fortunately didn't attend, was held during a union dispute and pickets prevented visitors from entering. The exhibitors got no compensation.

If your product is an item that is suitable for high street retailing you might like to ask the local privately owned toy shop, ironmonger, bookshop, whichever is appropriate, for some shelf space to gather sales data before trying to get it taken up by one of the chain stores. A few years ago, there was a company, here in Sussex, which had the apparently good idea of setting up stores with the express purpose of providing outlets for trialling new products or shifting remaindered stock. They thought they'd have it both ways: charge the supplier for renting shelf space *and* take a proportion of the profit from sales. Unfortunately the reality was they ended up with stores full of slow moving or non-moving lines. Browsing punters there were aplenty due to the high street locations of their 'Factory Gate' shops, but the tills hardly ever rang. They became a sort of permanent exhibition for failed products and soon went bust; just another example of how business is, like natural selection, a struggle for existence with winners and losers.

If people don't exactly fall in love with your product immediately, don't automatically become disheartened. Reconsider the product's qualities. Does it serve its purpose? Is it at the right price? Do enough people desire it? What about the packaging? How many of the ideal product criteria does it hit?

Even the most astute businessmen can have off days. Take the publishers Dorling Kindersley for example. I met Peter Kindersley in 1997 when I got an appointment (that in itself was an achievement) to propose an idea for grounding their 'lexigraphic design' of expository texts in pupil based research. I'd just done a similar exercise based on Usborne's books, DK's big rival, for my MA dissertation and I gave him a copy. He obviously liked the idea because DK subsequently involved the University of Reading in just such a project (but not me unfortunately) and their books have evolved appropriately. However, more recently, the company produced books to coincide with the relaunch of the Star Wars series of movies (Revenge of the Sith) and got disastrous results. The BBC reported as follows:

"British publisher Dorling Kindersley (DK) has seen its profits plunge after spectacularly over-estimating demand for Star Wars books. The company, whose shares almost halved in early trading, said it expected to make a £25m loss in the last six months of 1999. DK said it had sold just three million of the thirteen million

Star Wars books it had printed. It announced that chief executive James Middlehurst had left the company and that it was reviewing its strategic options with advisers."

Peter Kindersley has now retired from publishing and has appeared on the TV news described as a farmer. Since established companies like that can make mistakes you should not feel too depressed if your early attempts do not immediately hit the right button. After all, you're a beginner; it would be an amazing stroke of luck if the first product you present to the market turned out to be a winner. So just think about what you have learned, modify your strategy, and try again.

Once you have established that people *do* like your product and *are* prepared to pay money for it, you might want to exhibit it before the retailers and distributors. There are big international trade shows, such as 'Paperworld', which attract high street store buyers from many countries. Beware though, these shows are attended by predatory business men looking for any opening, including the opportunity to copy your product or challenge your patent. Don't give out any free samples or even sell single products (too easy to copy), just take orders for cartons or, preferably, container loads.

How much will they pay for your product?

Pricing may be a matter of 'trying it on' to see what reaction you get. Start high; when the customers blanch and walk away you know you are asking too much. Don't forget people often need a reason to buy *now*. They may like the product *and* the price, but want to keep their money for a beer later. You have to make them a temporary offer so they think, 'This is my chance, if I don't buy now I will miss the opportunity.' In my case that meant a 'Show price'. 'You can have all four colours today for £8, normal price £9.99 plus £1.99 p & p.' 'We've knocked £1 off the price and you'll save a further £1.99 post and packing by collecting them today.' Or you might try the most popular offer of the moment: BOGOF - buy one get one free. I'm sure you will be able to come up with something suitable.

Commoditisation: A commodity is something which is produced in large quantities and which has a tradable value. Unfortunately for the businessman, almost everything has recently become more 'commoditised' due to globalisation reflecting manufacturing costs in China - even the prostitutes are complaining that they are expected to perform for £5!

HOW SHOULD YOU EXHIBIT YOUR PRODUCT?

Exhibiting is an art and a business in itself. The first show I went to was in the Business Design Centre in Islington and I didn't hire any lights for the stand. I knew the venue had a glass roof and I thought I'd save money. On the second day the sky was covered in thunder clouds and my stand looked like the poor neighbour.

Nobody had plasma screens in those days so it was easier to compete for attention, but you do need to make your stand attractive with big coloured pictures and very few, very large words. If you have some intriguing shiny tinted products in a glass bowl on the table people will stop and look out of curiosity. Don't forget to make a Special Offer on them.

You will have to decide how best to present your product to a test group of consumers and you will find that the public is a hard master to please; people usually want something different from what you are offering. When I started selling the early version of the Wordwiza reading tool, I'd made a great big one for demonstration purposes which was exhibited on the back wall of the stand. Lots of teachers at the Education Shows asked if they could buy that one! This revealed that they didn't understand two things. Firstly, modern teaching is mostly about empowering children to solve problems for themselves, not the didactic experience of yesteryear, and secondly, there are about thirty times fewer teachers than pupils so the demand for a pedagogue version is too small to be worth tooling up for. Maybe we'll make one for the world market sometime.

I also got comments about the reading tool's colour. I'd made the first ones in transparent yellow because that was the colour of the first overlays for dyslexics. I soon had requests for them in blue so, ok, I made some in blue and took them to the next show. That's where I got asked for pink! I made some in pink and got demands for green. We have drawn the line at four colours despite being asked since for purple, brown and grey.

Movement is a big attraction; the United Kingdom's Lord Chamberlain knew this back in the nineteen fifties when he prohibited dancing by the nudes at the Windmill theatre. A friend of mine made a demonstrator for his product which opened it and slid the contents out repeatedly, as if by an unseen hand. Passers-by stopped to watch. Even then you can't please all the public. Some visitors didn't realise it was a demonstration model and complained that the versions on the display counter weren't working!

Then there are the people who staff your stand. The British public is a suspicious consumerate with a resistance to sales people; they fear they might be persuaded to buy something that they will later regret. There are three things you can do to try to counter this. The first option is to get attractive staff. Babes and hunks that people will want to approach! Take a look at the other stands; don't the staff look good? All bright eyed and bushy tailed! Salespeople know that the first thing they have to sell is *themselves*, so they make an effort to appear attractive. Now take a look at the show visitors: what a dowdy lot!

You don't want anyone too gorgeous though because confident folks will just waste time trying to chat them up, and shy people would be intimidated because they might feel they aren't in the same league. Let's face it, you want shy people to spend their money with you just as much as confident ones. Alternatively you can have ugly staff with a big plasma screen running demos and just use them to take money thereby reducing the public's exposure to their ugliness!

Another way, which works well if you have a small stand, is to leave it apparently unattended; people will feel it is safe to take a look at your products and that's when you appear from across the aisle where you have been skulking and say, 'Would you like me to demonstrate that for you?' This works even better if there are two of you. One can demonstrate to the other, who appears to be a 'customer'; a crowd will soon develop! Remember not to both wear the same company uniform!

"I'll do you then you can do me."

If you're disappointed with the attention your stand is getting why not try doing a 'Show and Tell'. You can buttonhole passers-by but, better than that, many exhibitions provide a little 'Come along' theatre area where the show visitors can sit and enjoy a performance promoting a product or service. Book a time slot and prepare an entertaining PowerPoint presentation. Don't forget to publicise this event.

The Pitch

You remember I said earlier that if you can't get the benefits of your product across in 30 seconds you probably won't be able to sell it? In America they refer to this as the 'elevator pitch' because the average elevator journey in an office tower block is one minute. The theory is that the brief ride in the lift is an opportunity to 'network' with your fellow passengers. So you should be able to deliver your USP (unique selling point), drop in the name of someone famous who recommends your product or is your ideal client, mention a 'strapline' that makes you memorable, and convey the company website address; all without gabbling!

In my case the elevator scenario happens at an Education Show and the pitch is made easier because our products sell themselves. For example, I just put an Easyriter pencil in the person's hand and say, *"Try holding that, what do you think?"* They remark, *"Very comfortable"* or something like that. I then say, *"Now try and hold it in a peculiar grip like children sometimes do."* They invariably say, *"You can't! It's uncomfortable unless you hold it correctly!"* I reinforce that concept by saying our 'strapline': *"Exactly! It's the pencil that shows you how to hold it!"* Then I do the 'namedrop': *"It's recommended by the Daily Telegraph and the Daily Mail. You can buy them from our website: www.ideasun.com – see - it's written on the flat side."* If they are still showing interest I can talk about how it doesn't cause writer's cramp or make dents in the flesh of the forefinger and thumb, it won't roll off their desk and the plastic lead won't shatter when you drop it. I can go on to tell them we are developing a pen which will be ideal for exam students who have to write for two hours or more.

The elevator pitch!

Having never had a single lesson on sales technique, I may not be the best person to advise you about it but I did start selling at age twelve and I have a hypothesis. It's that you should try to reduce any barriers between you and your 'mark' as much as possible. You only have a few seconds to bond with them. Use eye contact. If they are standing up, you should be standing up. If their nametag says they are a technical person, talk technical to them and so on. For example, if I'm pitching to an occupational therapist I talk about tensed adductor muscles and damage to the distal phalange of the second digit. If she's a teaching assistant who might hate me for showing off my pedantic knowledge, I use expressions like 'writer's cramp' and 'finger dents' instead.

Product Theft

If you are putting desirable tactile products into people's hands or leaving them on the counter to attract attention, you are bound to have some stolen.

Even a big notice announcing the price won't prevent people hoping to sneak away with a 'free sample'. Then there are the unashamed scroungers who blatantly say, *"If I could just take one back to show the others..."*

At an exhibition in Orlando recently, I was dealing with some customers when a bloke walked by and blatantly helped himself to a Wordwiza. He was talking into his mobile phone as he walked on so I didn't accost him, I simply thought, he's going to get home and say to his folks, "What do you think this is?" He hadn't seen the demonstration; it was just a total waste. Remember, you are running a business; if you give your stock away you will go bust. Point out that the Show organisers don't give away the stands and it has cost you hundreds of pounds just to be there, not including hotel accommodation and transport, etc. And you are making a special offer on your products which means they're getting a good deal... But for the small amount of stock that does go missing you'll just have to hope it turns out to be seed corn for a big harvest.

Testimonials

Trial marketing can also produce testimonials; you can ask people to write down their impressions of your products while they are at your stand. An endorsement from an obviously independent source is invaluable for giving your product credibility. In fact, the more you can get the better it will be. Some will say really complimentary things and others will be from highly qualified people with much gravitas. You can pull quotes from the really choice ones for your publicity. Always keep the originals in case you need verification when trying to borrow money from a bank.

If only I'd had a video camera at the Education Shows, I could have recorded lots of 'Wow!' reactions from visitors to my stand. Groups of women teachers with their jaws dropped, murmuring to each other about my amazing products. I must get that together one day... Until then, I'll just carry on looking up at an imaginary spot on the ceiling and calling out, *"Did you get that, cameraman? Zoom in on their faces!"*

Sending out free samples in return for a review is also good for getting testimonials. I did this with the Easyriter pencil by putting a message on the web-based SENCo forum (SENCo = Special Educational Needs Co-ordinators) offering freebies to the first ten to reply by email with their school's address. Over thirty responded and I sent samples to them all including addresses in Dubai and Australia. I got some great comments which I've put on our website. I also got a network of recommenders of my products and lots of sales by 'Word of Mouth' – the best and cheapest promotion method possible. And I've since made sales to Dubai and Australia.

A Customer Database

Since my free sample distribution was done to email contacts, I've been able to keep in touch with the reviewers and some of them are eager to test our next product. You can also ask visitors to your website to register their email address in order to be notified of developments.

Building a database of satisfied customers is like having a hotline to perfect clients. You know they like you because they've already bought your products. An announcement to them is a rifle shot, that is, it's likely to hit the target and result in a sale. Advertising to the general public is a blunderbuss by comparison.

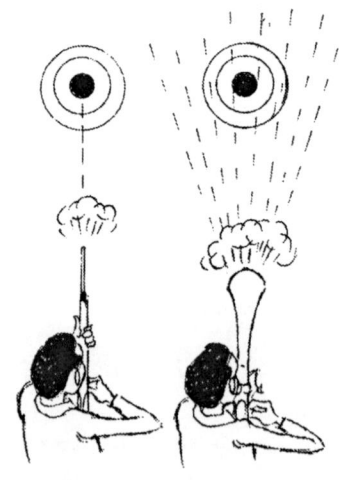

MANUFACTURING

Don't do this! No, seriously, don't do it unless:

- A) It costs next to nothing or
- B) You have already got promises to purchase large quantities of product or
- C) Someone else is paying for it – e.g. a grant

It is a bit of a Catch 22 trying to sell something that isn't available yet, but the last thing you want is capital locked up in boxes of unsold stock hanging around deteriorating for ages, so get a prototype made and see if you can get indications of intentions to order.

Manufacture it yourself?

We haven't gone this way (yet), but some people have made a success of it, including James Dyson. It's an entirely different game from inventing, with a workforce to manage, premises and equipment to maintain and long hours of work; a bit like being an employee, but of your own company. You've got to set up a factory, buy expensive machines and learn how to use them, employ and train staff, order materials and overcome problems of supply, handling and delivery. There is a huge capital outlay and regular outgoings in the form of wages to pay, heating bills, business rates, raw material supplies, etc.

Historically, inventors *had* to do it this way because they were also inventing manufacturing as an activity. The twenty first century is better organised. Sub-contractors are available with equipment and expertise to carry out all the processes that you are a novice at and, if you simply copy what they do, you may be wasting your inventing talent by filling your life with mundane repetitive operations. Also, they can often do your work cheaper than you could because they have several customers so their machines are seldom idle. Even the big car manufacturers use sub-contractors these days. I'm not the manager type so I don't recommend doing your own manufacturing until you've made so much money you can buy your sub-contractor's business.

Licensing?

An invention tends to become the inventor's 'baby'.

Intellectual property is an even more *personal* possession than material items. He or she will resist giving it away or sharing part of it. For that reason many inventors are reluctant to allow someone else to manufacture their product in return for a royalty. That's probably a wise attitude because, proportionally, there's very little money in licensing. Unless you have a really stunningly good idea for improving an existing product, in which case the vested interest might offer you a million to go away and stop spoiling their game, you will be lucky to get as much as 4% and, more likely, will receive only 1% of the final price.

> *The executives of big corporations like to hunt in packs and surround their small prey, frightening it one by one before baring their fangs and all pouncing together. One of them says, 'We'll give you 5% of the first 100,000 units then 2% on the next 100,000...'*
>
> **James Dyson**
> **'Against the Odds'**

There is no fairness or moral justice in business. It's the one area where might is still right. Your licensees might be, for example, an internationally recognized kitchenware firm while you are Mr Man-In-The-Street with a mortgage. Who will have the power in a negotiation? You've guessed it, so prepare to get ripped off. The reason behind it is they are established in the old Victorian structures of commerce. They are links in a supply chain where every link feeds off the product as it moves from manufacturer to consumer. Remember the small slice of the pie that high street retailers are prepared to let the manufacturer have? Well, your licensee will be doing business with them so he might only get that 12% of the action. Unsurprisingly he doesn't want to pay you many of those twelve for permission to use your invention.

Mark Sheahan FRSA, inventor of the Squeezeopen plastic lid for containers, has years of experience of pursuing the licensing route. Mark has invested £750,000, mostly of his own money, in developing his excellent device for which he has successfully obtained five licensing agreements and is working on three more. He tells how you will have to designate areas of the world to sell licenses for, obtain a signing fee and a royalty, and warns that you must make sure to have termination clauses in case your licensee goes bust or gets taken over. He advises that the inventor is the wrong person to close a deal and recommends that you get a licensing solicitor.

Visit his website (www.squeezeopen.com) for the full story including how he was 'mugged in broad daylight' by a copyist.

> *To be honest I've given up on England. I got Silgan Plastics within five weeks of going to the States, whereas I've been knocking on the doors in England for about two years. I think there are some forward-thinking progressive companies out there (in the U.S.A.). Unfortunately, Brits want proven history and don't want to jeopardize their safe job by taking risks; it's a shame.*
>
> **Mark Sheahan FRSA**

Sub-contracting?

Finding a manufacturer for your invention can be a nightmare. For a start, the whole manufacturing sector is currently transferring to India and the Far East. We are living in strange times; the infancy of globalisation. Here in the West, right now, we are enjoying a standard of living much higher than people in developing countries. Consequently labour is much cheaper in those parts of the world and therefore it's more economic to manufacture in locations like China and Indonesia. In fact, if a manufacturer doesn't take this opportunity, he will risk going out of business because his competitors, who *are* sourcing that way, will be able to undercut his price. James Dyson has just had to move his cyclone vacuum cleaner manufacturing plant from Wiltshire in England to Indonesia for this very reason. If it happens to you, people will accuse you of not being patriotic but you can't change the world single-handedly, you just have to live in it. In a global world, what's the point of nations?

Naturally, the workers in developing countries aspire to raise their standard of living and will study hard and work hard to achieve this aim. They will gradually become able to enjoy the fruits of their labours when local infra-structure, such as roads and electricity distribution, improves enough to provide them with opportunities to spend their wages on luxury goods like cars and TVs. This is the expanding market I alluded to earlier. In China, some factory workers are already beginning to realise their power and are withdrawing their labour to demand more pay.

Meanwhile, in order to maintain their economic status, the developed countries have had to focus more on service industries like design, insurance and TV programme production (media now forms 17% of UK earnings!) or high tech manufacturing with heavy development investment such as the production of aircraft, satellites and pharmaceuticals.

Eventually these global disparities may equalise, which would mean that the cost of labour in the developing countries might rise to a level which could make it competitive to manufacture products in the West once again but, at the same time, demand might plummet due to the increased cost of those very goods and the, by then, saturation of the market. Different products would be affected in different ways; cutting edge goods will usually succeed but anything old-fashioned will struggle. An alternative scenario might have China becoming the planet's major economic nation and the next world superpower. Such predictions are beyond the scope of this book but the relevance for an inventor in the early twenty first century is this: it's going to be hard to find a suitable manufacturer.

An inventor in the Western world has a choice between discovering a factory in a diminishing local supply, or overcoming language, culture and distance barriers to find one overseas. There are agencies that try to help with the latter; however, in my experience, nothing is ever easy. For example, India may have lots of well trained, English speaking workers who could produce your product economically, but the country is overrun with form filling bureaucrats that slow progress down to a crawl. You have to be prepared to live a very long life to get anything done. I'm still waiting for a quote that I requested from an Indian agent in early February 2005.

So, one very important consideration when selecting a manufacturer is the location. This is not only because of the low cost of labour in developing countries and the risk of copying, but also because some places come with money attached. Development areas in the UK have all sorts of grants and advantages. Developing countries attract International Monetary Fund finance and, if you can collaborate with another small company in a different member state of the European Union, you may be entitled to enterprise grants as long as you like filling in forms and hitting targets.

Another significant factor is the nature of your product. If it is one piece of stuff that comes straight out of a machine it won't matter so much where you have it made. You'll probably be paying the same price for raw material wherever your machine is sited, since commodities like plastic are traded on the world market. If your subcontractor is in the West there will be higher costs for a machine operator and capital investment in the factory etc, but these will be small components when spread over a long production run. However, if your gadget is in several parts that have to be assembled and packaged you'll need to take advantage of the low cost of labour in developing countries. You don't have to get involved with 'sweat shops'; there are plenty of well run factories offering decent pay within the local economy which, because of exchange rates between different currencies, is still a bargain price in the West. There are also introduction agencies and

Western based companies with their own overseas manufacturing facilities to help you meet your requirements.

When I was searching for a manufacturer for our Easyriter pencil I discovered that there are only three pencil makers in the UK. One makes high quality traditional wooden pencils and crayons for the Fine Art market; another is a Welsh branch of a German pencil company and the last one is a tiny firm which recycles plastic vending cups into pencils by a co-extrusion process. I checked out two of them. The artist pencil company said they could make our pencil in wood but it would mean a large investment, they wouldn't be able to print on them and they weren't interested at the moment. After a lot of persuading, the recycling firm said they'd have a go. A year and £1000 of my money later they had produced a few samples of unsatisfactory quality. It was such a pity because I had shown those early production versions at Education Shows and got an ecstatic response. When told that they weren't ready for sale yet teachers said, *"But I want one now!"* so I broke a piece off and let them have it.

It became obvious that this manufacturer had other priorities. I asked what his schedule was and he said they couldn't see their way to doing much work on the project for about another year! My patent application was ticking away so I gave up on them and sought out experienced plastic extrusion companies. Luckily I soon found one that was willing to take on the job. They had to buy a new machine, and then part of it broke under the pressure. After repairing it and conducting a series of trials of different materials they finally found a combination that worked; we are indebted to them for their determined co-operation. The moral of that story is, **if you can't find a specialist manufacturer for your product go for the most experienced generalist.** If only they had been able to do it within their original quote and not hike the price up by 250%! Will we stay with them? It will depend on what quotes we can get when we next need stock...

Even when you get a good manufacturer there are still traps to step into. Having successfully used the extrusion process to make pencils I thought I would apply this technique to pens. An extrusion comes squirting out of a machine quite fast so it seemed like a good idea. I had the extrusion company tool up to make a pen barrel to my design and gave some samples of them to a CAD firm to design the pen tip which receives the ballpoint refill and the stopper for the top. We worked together to come up with satisfactory shapes to go to the injection moulders and had a stereo lithograph made. That was when we discovered that the extruded barrel had dimensional imperfections. At first I thought it was of such variable measurement that it was next to useless and had to consider making a barrel in one piece with the tip by injection moulding. At least that would cut one operation from the assembly process. By the following day everything had changed again. The injection moulders said they couldn't produce a combined barrel and tip long enough with such a narrow core, and the extruders said they could improve their production of the barrel so we were back to plan A. Since then we've had so many demands for the pen from people who have bought the pencil that we've decided to abandon the tactic of having a few made by local manufacturers for trial marketing and go straight to overseas mass production. This has necessitated a total redesign and we've decided injection moulding is the way forward after all. That earlier effort was wasted. Just remember, nothing is ever easy!

Getting acceptable quality from a manufacturer is another potential battle. You remember the pirate who produced the counterfeit reading tool? He fell into the trap of making a cheap copy in brittle material that broke as soon as you looked at it. We had been through two production runs of different plastic materials before we found one that could resist the attacks of children. Unsurprisingly his fragile copy didn't sell for him, but he gave lots away, and probably spoiled our market by associating *our* product concept with *his* inferior quality.

There's been a lot of talk about 'just in time' manufacturing. The theory is you save warehousing costs and interest on borrowed capital by only making goods to order. If it's mail order by credit card you even get paid up front! It really depends on your customers being prepared to wait for fulfilment, and some will. In the US, the public has been trained by over a century of mail order catalogues like Sears Roebuck, which was launched in 1888. All those farmers in vast areas of land with no shops nearby have an expectation of one month's delivery.

Certain products can be delayed more than others. If you desperately want a Morgan sports car with a chassis hand-made in ash wood, then you must accept that you have to join an eighteen month waiting list but, if you want the latest hit recording, you want it now. Mind you, with downloading of MP3s becoming the norm nothing has to be manufactured and warehoused anyway... The same thing is happening to the book industry – digital printing enables books to be produced singly to order. We live in a changing world. Publishing houses must be trembling, especially as the flexible screen is coming out soon – imagine an iPod for the written word...

'Just in time' is still difficult for many new physical products: you really need a predictable sales stream to be able to manufacture to order and, if you are using a sub-contracting manufacturer he will want you to buy a quantity which will probably require warehousing. However, if you've followed my advice about making your product small and light you may be able to use your garage as a warehouse. Don't tell your property insurers – they are happy for you to keep a car in there, with a tank full of gas, but boxes of plastic goods – not on your nelly!

Do it yourself overseas?

This entails setting up a 'free-zone' in a developing country. There are pros and cons as usual, and it varies from country to country. Let's take the West African state of Ghana as an example, just because I know a bit about that country since the second Mrs Richards comes from there: first you have to find some suitably secure premises with a twelve foot high fence and fill in lots of forms guaranteeing that at least 70% of your production will be exported, etc. You need to show equity of $50,000 unless you have a local partner when it is only $10,000. Then you have to pay a fee to the government of $1000 (plus $1000 annually) and import your machines, raw materials, train your workforce, etc. Once you have set this up you don't pay local taxes for ten years and can import cars and well, anything really, duty-free.

It may not be the solution to the initial manufacturing problem but definitely something to review every time you want to ramp up production; especially if the International Monetary Fund will provide capital funding...

FINISH AND PACKAGING

It's such a shame for the serious inventor to have to come to terms with the fact that the container it's in can be more important and valuable than his newly invented gadget. It's as if the box is more nourishing than the cornflakes! Think of the upmarket cigars that became popular in the nineteen fifties. They were packaged in a torpedo shaped spun aluminium case with a screw cap. That was the state of the art at the time. Aluminium had only recently ceased to be a semi-precious metal because the Hall-Heroult process of electrolytic extraction had become economical due to the construction of large plants beside hydroelectric power stations in the early nineteen hundreds, and output was climbing. What did they use this wonderful high tech packaging for? As the wrapping for a cylinder of dried leaves intended for burning! But the cigar could now command a much higher price...

Plastics have altered the customer's perception of finish and packaging enormously. We used to be satisfied with fish and chips served in yesterday's newspaper or sugar weighed out and wrapped, in front of our eyes, in a cone of blue paper. Now we expect a burger to come in a polystyrene or cardboard box with a full colour label depicting the deliciously photographed contents. Shrink wrapping or blister packs are the norm for everything from kippers to screws. We don't even have to take a wicker shopping basket with us like we once would have; we can be sure the shop will give us a polythene bag, publicising their name, to carry our purchases in.

Products can be finished in any effect from metallic silver, through velvet textured mock leopard skin, to holographic pictures or diffraction gratings: opalescent blue one minute and pink the next. These finishes can be applied to almost any material by techniques such as ink jet printing, powder coating or shrink wrapping. High quality surfaces on plastic, such as transparent fluorescent colour with an edge glow or metallic speckle can be obtained straight from the mould.

Unfortunately this level of capability has led consumers to expect such a high class of finish and packaging nowadays that it is very hard to get them to accept a home-made product unless it is organic marmalade, and that would probably fall foul of the Environmental Health Department of the local Council. It's difficult to get a representative customer response from a cottage industry level of presentation any more. So the poor cash strapped inventor has even got to tool up and machine manufacture his prototypes for trial marketing. This means you have to make a big financial wager early on in the game and you will likely get involved with expensive retooling when improvements suggest themselves after the first production run.

Your packaging ought to be noticeably different from your competitors and, if possible, it should reveal or reflect your product's unique sales point. We made the Easyriter pen carton triangular for this reason. It's the shape that's never done Toblerone any harm! No other pen is packaged in a transparent triangular box. And we cut a hole in it so customers can feel the special profile before purchase.

MARKETING YOUR FINISHED PRODUCT

Marketing is about perception. It's about becoming famous and establishing the notion of a trustworthy company with desirable products in the eyes of the consumer. You're going to be asking some of your customers for their credit card details; they have to believe that, in a 'customer not present' transaction, you will only take the amount you've told them from their account. So it's better *not* to reveal that, on the end of the telephone, it's just one man and his dog. In *our* early days there was one man (me) and *no dog!*

You have to create an impression and *you have to become known*. **This is enormously difficult and, therefore, enormously expensive.** In comparison, inventing is easy; manufacturing is difficult but possible, while marketing is verging on the impossible. It's a black art. Did you know that 1% is counted as a typical response from leaflets? You know all that junk mail that comes through your front door or falls out of your newspaper? Ninety nine per cent of it is junk to the sender too! I realise this is not very comforting information. Just think how much cheaper those mail order shirts could be if they didn't waste 99% of their marketing budget on failed mail shots. One small company told me they spend £250,000 a year on producing their catalogue. That's a huge commitment for the penniless inventor to have to consider and, let's face it, he's not really interested. The fun for him is in creating products.

In order to be successful, unless you have some secret giant killing technique, marketing has to be seriously expensive: Global Fund's 'Red' campaign to raise money for AIDs has been reported as costing $100m and as having produced a paltry $18m in donations. If true, it would have made much more sense to donate the entire marketing budget to charity!

The market is like the air. It consists of billions of atoms that represent the consumers. Every consumer/atom has a nucleus which is the wallet or control centre for finances.

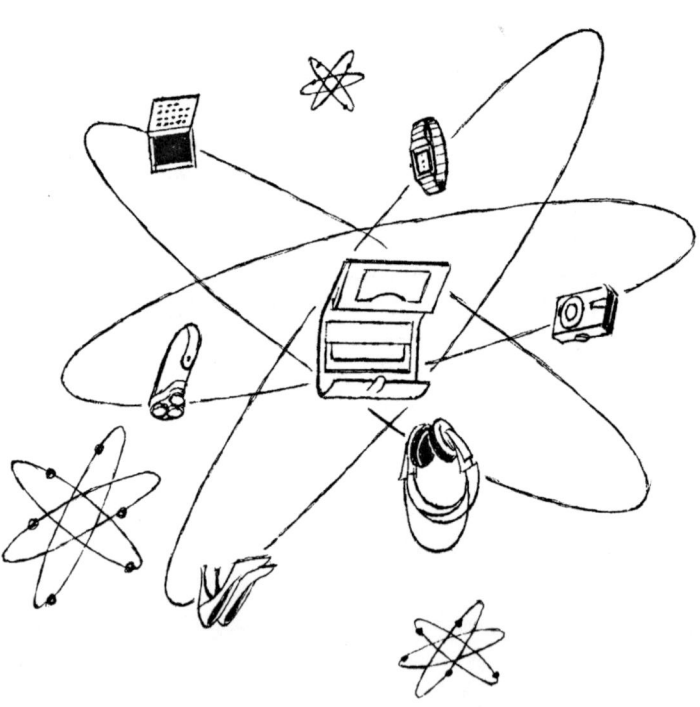

Each nucleus is surrounded by swirling electrons that represent the products and services competing for the 'spending money' (disposable income). That list of things I used earlier will do: cosmetics, confectionery, hairdressing, fashion garments, jewellery, tattoos, gas guzzling sports utility vehicles, computer games, football match tickets, alcohol, etc. You, as the seller, have to cut through all the existing competition of time consuming interests and demands for expenditure, in order to get pole position admission to the purse strings.

Before you even get that far, you will have difficulty because these 'consumer atoms' are not necessarily easily accessible. Just like the real atmosphere, there are high pressure zones where they are packed tightly together; this is a boom location or time. Other areas are very sparsely populated; this is a depression period or area of deprivation. Make no mistake - selling will be difficult under all conditions and therefore expensive to achieve.

Marketing makes use of the media; the means of communication for getting your message out to the potential customer. There's the press, leaflets, posters, radio, television and that new boy, the internet. Don't forget you can use them in combination: a multimedia campaign.

Buy me now!

We are all guilty of procrastination. Unless there is a danger that we might miss an opportunity, there is no reason to act in haste. I've lived near Arundel Castle for over thirty years and I've never taken the tour round it since I can do that any time. When I lived on the Isle of Wight I never went to Queen Victoria's favourite residence, Osborne House, because it was so easy to get to there was never any need to rush. As a result, I've still never been. Yet, when we visited San Francisco we went on the boat trip around the bay while we were there because we didn't want to miss the chance. I bet you live near something that attracts sightseers but has never attracted locals like you!

So for that reason, most items that are on sale today, including essentials like bread, come with some message of psychological imperative exhorting you to purchase them at this very moment. A product has got to be promoted as Special Offer! Buy One Get One Free! First Ten Purchasers get a Free iPod! Last Few Available! Sale Price! Extra Points on your Clubcard! You will be expected to come up with some sales ploy of this sort.

The Press

I've usually found that advertising in the press is unproductive. It's such a limited medium. There's no sound or moving pictures and reading masses of text is going out of fashion. Free newspapers often go straight in the bin. Even the paid for newspapers just get scanned; a guy might read the sport pages and ignore the rest. Girls may focus on the gossip columns and horoscopes; nothing else. Ok, that's generalising but, since marketing costs a massive amount of money, you want to be sure to hit the target so stereotypes suddenly have value!

The best result we got from a newspaper was when the Easyriter pencil got mentioned by the Daily Telegraph's education agony uncle (see www.ideasun.com). It was just four lines

Articles for Sale
Commode £5
Stool £2
(Ask a nurse to explain...)

in a single column but they gave the right phone number and spelt the website address correctly. Our phone line went crazy. Third party endorsement like that overcomes the 'automatic disbelief in an advert' response.

Whatever the medium, it is always better to advertise sneakily so that isn't obviously an advert. There are ways of doing this. In printed media the technique is to make an advert look like an article or a piece of editorial. It's called 'advertorial'. If you can get a recommendation from an expert who is perceived as independent, such as a journalist, that will be wonderful. You will probably get your greatest response from something like that. So send out press releases about any news you are creating, like a product launch. Articles with good quality photographs are best. Write in the style of the publication you're submitting it to, sometimes the lazy beggars print it verbatim!

Glossy magazines may work if you have a huge budget and can buy a whole page in photographic quality colour like the ads for Jaguar, or if you are modestly going in the classifieds – there is a small audience of, mostly older, people with lots of leisure time who read those.

Some newspaper editors are sticklers for procedure and won't include details that would enable readers to buy from you, possibly because they want you to pay for an advert. There is a way around this however. You get someone with an address which has no connection to your company to write to the Letters to the Editor page saying, *"I read your excellent article about such and such a gadget, where can I get one?"* and the chances are you'll get another mention. It will depend on how desperate they are to fill space. On one occasion we got another article with a picture and the next time a different paper simply wrote back giving our contact details to the enquirer who was my cousin! You do need to know the publication's demotic though. If your product appeals to grannies you could waste a lot of your time trying to get mentioned in a teen girls' mag.

Warning: Scamming Publishers

Watch out for scamming publishers; they have various techniques. Some try to sell you space in a publication which has a massive 160 pages and a print run of only 3000. Even if all the readers scrutinise all the pages (what's the chances of that?) your advert will only hit a tiny number of people. Then there are the 'directory' merchants; they want you to pay for an entry in an obscure publication that you've never heard of. If *you've* never heard of it, the likelihood is neither has anyone else.

Some use the name drop technique. This is where they call their publication after something with a high profile and impressive reputation like Parliament or Senate House and say they are distributing copies to all the Members of Parliament or Senators to make it sound as though they have official backing and therefore you'd be foolish to turn down this opportunity. They don't mention that MPs and Senators get swamped with more important papers to read and have no time for speculative advertising magazines. Others send you an invoice for publicity you haven't ordered in the hope that your finance officer will be on autopilot and will pay without question. Or they send a solicitation that *looks like an invoice* until you read the small print. Don't forget, the job of these deceivers is to fish for and hook their victim – you.

Leaflets and Posters

These time honoured promotion strategies still work for some businesses. If you're opening a new restaurant or takeaway you know your customers are within a small radius and you can put your menu through every door. You can also put up posters in shop windows or village notice boards. But for an invention or a new product it's a different story. Your potential customers may not be so geographically confined. You may want to contact them at work rather than at home. You might even consider telesales or cold calling. Ugh! Have you ever met anyone who welcomes those marketing strategies?

Radio

Radio is better in some respects; there's quite a listener-ship of motorists and housebound folk for some programmes and it can reinforce a multimedia campaign. Radio interviews though are too transitory; you can blink and miss them and they are rarely repeated, except for snippets on the news, although the 'listen again' facility on the BBC website is useful. The main problem is the old 'juggling on radio' one; you have to do verbal descriptions of new products. You can hear me doing this in a BBC radio interview on our website (www.ideasun.com).

Television

The most powerful medium at the moment is definitely television. However, increasingly sophisticated, and therefore sceptical TV viewers are becoming aware of the bogus nature of commercials. Is anyone fooled into buying shampoos because they claim to contain 'Nutrillium' or 'micro oils'? I can imagine Scotty of Star Trek saying, *"Och, the Nutrillium crystals canna take it any longer, Captain!"* And, micro oils? Very small oils? If you make the molecules of an oil smaller they become a gas; it's just absurd! It reveals an assumption by the marketers about the intelligence of the viewers that is extremely insulting. A few years ago I predicted that the next generation of products would follow the 'Small is Beautiful' trend started by economics author E F Schumacher and commercials would soon feature the words, 'nano' and 'pica' because they mean even smaller than 'micro'. Sure enough, it's already started. I've even bought iPod nanos for my wife and myself.

You may remember when 'Turbo' was the buzz word and we had such ridiculous things as 'Turbo' hairdryers and vacuum cleaners. Anyone who knew that the word was coined to mean a car supercharger driven by the exhaust gases was entitled to smile cynically. We even had a button marked 'Turbo' on our first desktop computer!

The Turbo Corkscrew!

When were you last inspired to buy something because of a TV ad? Were you ever? They can't *all* be aimed at the gullible or children, can they? And their days may be numbered. When we all have hard drive recorders we might watch the programmes and skip the ads. There is already a recorder, initially being marketed in Australia, which you can set to do this for you automatically.

If the commercial break does disappear we'll probably see an enormous increase in 'Product Placement'. This is when companies pay to have the camera linger over their product *during the programme*. The Dyson vacuum cleaner has already done guest appearances in 'Friends' and 'Will and Grace' and the Segway personal transporter has been on 'Frasier'. So you'll get to see the Desperate Housewife, slightly out of focus, across the table from an enormous, sharp image of a Heinz tomato sauce bottle in the foreground. She finishes her lunch and goes out to get in her car. The camera shows a close up of the Chrysler badge on the wheel trim as she drives away.

This subliminal stuff will work much better because we won't realise that they are trying to influence us. The days of volume compressed ads simply shouting the virtues of a product like, *"Don't just clean it. Oxyclean it!"* may soon be history, and few will be sorry. Or maybe they will just move them to the shopping channels where their fans (!) can watch them.

The latest TV ads, seen here in the UK, have begun to move in the product placement direction. They are like mini-biopics about the commercial's leading man, showing his lifestyle and his weaknesses so you are wondering what product they are pushing and then, in the last few frames, they show his more successful competitor getting into a Volkswagen car. I haven't seen these in the USA yet. Being a city based country, most of the ads are low tech local ones featuring simple coloured screens with text and a voiceover.

The TV equivalent of advertorial is the 'infomercial'. This is a programme focused on reviewing products. There are lots of motoring examples and several showing the latest gadgets. To some extent the magazine style programmes hosted by a man and a woman sitting on a couch fall into this bracket because they might interview an inventor about his latest gizmo or demonstrate a new kitchen toy in the cookery slot.

Unfortunately it's very difficult to get on TV unless you're a producer's daughter! It's much easier to get coverage in newspapers and on radio. Even if you think you have a significant news item the chances are that something more sensational will happen that day and take precedence. In my politically active days I staged a protest about sewage in sea water and was interviewed in front of the outside broadcast cameras. We had men dressed in wet suits with colourful windsurfing equipment and everything. When I watched the local news that night there was twenty minutes about Southampton football manager's surprise resignation then five seconds showing me walking down some steps just before the comical duck on a skateboard item!

However, there is no doubt that, if you can get your invention on TV somehow or other, you will experience a huge boost in sales because TV exposure (although not necessarily in the form of a commercial) works best of all the media. Even if you personally are too sophisticated or cynical to respond to adverts, you have to admit that television is such a powerful medium it has recently redefined 'celebrity' to mean undeservedly famous! One witty journalist has dubbed these newly prominent people 'sublebrities'!

Celebrity endorsement

Celebrity endorsement is another good way of combating consumer scepticism. It relies on the reputation of the star to overcome any lurking doubts in the genuine nature of the product and there is a whole range of 'A' list to 'C' list celebrities or even celebrity look-alikes you can hire to put your message across. Imagine having your product recommended by 'The Queen'! Beware of the new breed of 'sublebrities' who have become famous on the 'strength' of being vacuous though. Their endorsement may have a negative effect! Mind you, an ad which shows Jade Goody studying the Encyclopaedia Britannica might be good for the publishers! A need fulfilled! Post modern ironic or what!

Sir Richard Branson, a national hero and Virgin's own in-house celebrity, has to be admired for the way he has raised the profile of the Virgin brand by acts of 'derring do'. Maybe he did spend a bit on transatlantic speedboats and global circumnavigating balloons etc, but he has achieved acres of sustained coverage in all the media without paying for it, had some fun, and the 'Virgin' label has become internationally famous. It can be used to promote almost any product, except contraceptives...

... and so they called the condoms 'Mates', but everyone knows they were a Virgin product.

Sir Richard is presented by the media as an icon in the buccaneer mould who is synonymous with his brand and imparts an image of glamour. Hardly anyone knows who 'Mr Coca-Cola' is, and that firm has to spend millions of dollars claiming their product is 'the real thing', while Sir Richard is his company's own celebrity and requires no endorsement fee. Shrewd, eh?

Although many people realise that advertisements make bogus claims, when Sir Richard is associated with his products it's more like the word of mouth recommendation of an admired friend. The subliminal message is, 'You can be the glamorous and successful owner of a Caribbean island like me, if you use this product or service'. He doesn't have to say his stuff is good; you can see the quality in the enviable lifestyle. And now, *Virgin Galactic!*

The Internet

The latest marketing medium that we have to add to the list is the internet. With the expansion of broadband this is becoming a serious tool. Your website can have moving images and soundtracks and still download in a reasonable time. The difficulty is, as in every marketing initiative, attracting the interest of the customer. How do you get your website to be visited? High street shops talk about 'footfall' past the premises and theatres try to get 'bums on seats'; on the internet the equivalent is 'mouse clicks'. There are always millions of people surfing the internet and they are just a click away from your site. Trouble is, if they don't know it exists, how will they find it? Well, there are ways of directing them to your website when they are using a search engine like Google. You have to get up to the top of the search result list. It's complicated, expensive and too boring for this light read; just get your webmaster to do it for you.

Once you have got them to your website you need to clinch the sale which is about having a user friendly site with a shopping basket facility: 'e-commerce' as it's called. Strangely, many people still have an aversion to putting their credit card details into a website. They will happily phone me and give me their card number and expiry date thereby trusting me to take the correct amount from their account when I could easily withdraw more, but they don't like a computer to do it. If I was really wicked, I could buy things for myself over the telephone or the internet using their card and address details. I haven't tried pointing out that, when they buy from our website, it doesn't tell me their card number *and* they get to confirm the amount themselves, because I don't want to lose the sale by sounding patronising. We have provided a downloadable and printable order form, addressed ready to snail mail to us, for these technophobes.

You also need web surfers to visit again soon. This can be done by asking for their email address and sending them a message when you have something to say like the launch of a new product related to the one they bought, or by having some 'content'. Content turns a website from an online catalogue into a magazine. The latest thinking involves 'humanising' the website with narrative to increase the appeal, especially to women who like a good story. We are turning our website (www.ideasun.com) into something a bit like a computer game experience.

Hosting a message board, providing relevant news articles or free downloadable resources for your clients might make them into regular visitors. Children can be attracted by games or online competitions and, pay attention to this, youngsters are very important people. If you can get your company name pleasantly embedded in their memories when they are littl'uns you could have their custom for life; think Disney. The value of this may depend on the age range appeal of your products, of course, but even if you just supply kids' toys, they may remember your name later in life when they start to breed themselves.

Last year I said to my business partner, Richard James, *"Who, in their right minds, would produce a TV channel with no programmes, just the adverts?"* What I was thinking of was the fact that the internet is rather like that; a broadcast catalogue which has very little watch ability, except for the porn sites. The very next week they announced the launch of a TV channel that only shows adverts! It's for the ad freaks who love to watch their favourite commercials and learn all about them. Who'd have thought it?

However, the day will come when the content of websites approaches the attention grabbing ability of TV programmes. Either that or TV programmes will be dumbed down until they are completely ignorable like websites with no content. Cynics would say that some channels are nearly there. One way or another, the two technologies will certainly merge; you will download TV programmes *on demand* and watch them when it suits you. When websites can *captivate* an audience to the same extent that cinema and television already can, that's when the worldwide web will become a really powerful marketing force but the growth in ecommerce is already phenomenal.

I'm revising this book now and, since I wrote the above, 'pod casting' has actually begun and so have 'virals'. These are little videos that appear on Youtube, or similar websites, which people send to each other because they are funny or of some other interest. Imagine that! If you can hit the right button, people will distribute your video from pc to laptop or phone to phone for you at no cost! And, if you can appeal to your target market it's job done! How much cheaper and better aimed can marketing get?

And the very latest news is about Joost. A peer to peer video streaming software that will enable us to watch full screen live TV programmes on our computers and IM each other about what we are watching at the same time. We'll each be in our single occupant houses enjoying an 'en famille' experience over the internet! In future we won't have to wait until the coffee break in the office next morning to discuss what we saw last night!

I've got to mention Ebay. It seems you can sell anything on there including your latest invention provided you don't mind rubbing shoulders with all sorts of miscellaneous other products.

NB UK Internet sales have increased enormously in the last year (2006): 28.9%.

Marketing Companies

In order to attempt to do business at all you have to be an optimist and, since marketing is the most difficult aspect, if you are in the marketing business itself you have to be a super-optimist. That's why most marketing company employees are young. When you are young you imagine that you are invincible, possibly superhuman, and you can believe that all your prospects are good. Later in life, as you feel yourself being sucked towards the plughole of incontinence and dementia that you've already seen your elders spiral down, it's more difficult to maintain that kind of positive attitude!

I've been to marketing firms; the first thing they ask is, *"What's your budget?"* Listen matey, when I go to the greengrocer for a lettuce, I want to be told the price; I don't want to be asked, *"How much would you like to pay?"* I'm reminded of the times when I've phoned in an order for Chicken Tikka Masala and been asked by the Indian takeaway, *"Do you want all breast meat?"* I always say, *"No thanks"* because it's *always* all breast meat – check it out, there's no dark meat in it. What they're actually saying is, *"Would you like to pay more?"*

So, until the marketing boys get their act together and produce a menu of different levels of service with price tags on them, they won't get my business. Amazingly, they expect you to hand over a cheque for, say £300,000, and let them spend it as they see fit.

Value for money will be the last thing on their minds. For a start, it's not their money. Secondly they can't make any guarantees of success and, therefore, you can't sue them for incompetence so they have nothing to fear.

You tell me what strategies you have available, Mr Marketing Man, with the prices alongside, then I can choose whether to spend my money with you or go to the other firm that offered me the same service for less money. Don't give me that crap about things not being comparable. You have to submit expenses for your accounts don't you? So you must know what your costs are. Stick a modest profit margin on your costs and let's have a standard bill of fares. A builder can give an estimate for a project, why can't you? You've had it easy for far too long. Ok, the builder often underestimates and subsequently has to put the cost up...

Warning: Invention marketing companies

Once you have been granted a patent your name will appear on a register of patent holders and you will receive mail shots from companies offering to find you a manufacturer and market your invention for you. Some even advertise that they can help you get a patent in the first place. Beware; they are after your money! They charge you a small fortune for some general advice and send you a list of manufacturers or retailers: often inappropriate contacts. Many of these have been exposed as fraudulent and there are websites that help you to identify them: www.bpmlegal.com www.inventorfraud.com

The British Patent Office also has useful advice:
http://www.patent.gov.uk/patent/howtoapply/ipromoters.htm

Does Marketing work?

It's the ultimate hindsight activity; nobody knows until afterwards whether a marketing campaign will work. It's a shot in the dark. A 'poke and hope' in snooker jargon. Here's an example. Back in the days of black and white television there was a very evocative commercial for cigarettes. It was set at night-time and starred a hunky male actor in a raincoat walking dejectedly, on wet cobblestones, down a deserted street. The background music featured a harmonica played by that late, great, Anglicised American, Larry Adler. The melody lingered hauntingly on the minor intervals. The man came to a street lamp and stopped in its pool of light. Droplets fell from the brim of his trilby as he pulled out a cigarette and lit it. Immediately the tempo of the music picked up and the melody bounced jauntily over the notes of a major chord. The throaty baritone voice-over announced, *"You're never alone with a Strand."*

It was a mini masterpiece of cinematography. Everybody loved it; people remember it fifty years later because it made such an indelible impression on them. But it didn't sell the cigarettes. On so many levels it was a success, except at achieving its purpose. Perceived wisdom today suggests that, since smoking was a social activity, portraying it as a solitary pursuit just associated it with lonely losers and who would want to buy into that? Whatever the reason, for the manufacturers it was just a huge waste of money, but I bet the marketing people came out of it as earners…

Established manufacturers of famous products may be able to point to sales graphs going up during a TV ad campaign, but it's a very different story for anything *new*. You remember the dotcom collapse of the year 2000? It wasn't a failure of the products or services they were offering. Mostly they were just offering a lower price on the same stuff you could already buy elsewhere. It was a failure of *marketing*.

Let's take a look at a particular case: a dotcom company that followed the marketers' handbook to the letter. No, I'm not going to name them; I don't want to be sued. They researched everything, seeking the advice of focus groups. Out of a selection of trendy abstract words they came up with the company name. It was empathic. It was vital. It was a verb! They chose a colour: blue, like the sky. It was optimistic; redolent of happy days. The best experts designed a logo. Then they confidently swung into action with their media campaign. Television commercials were reinforced with high street posters and full page, full colour ads in magazines. All of this cost almost £40 million pounds. They were just haemorrhaging money and had to start laying off employees. Yet, to this day, even close relatives of the ex-employees don't know what they were trying to sell! Was it insurance? Dough nuts? Surgical appliances? Yes, the message got out that badly. I bet they wish they still had that £40m. I hope they think they got good value. Remember, no marketing person is ever going to say, *"Don't spend your money with me – I think it will be a waste of your resources"*

Under those conditions marketing simply doesn't work. You can't expect to change people's habits by a short to medium term advertising campaign. Look how many decades it took to educate the western public about the dangers of smoking, and hundreds of thousands of people were *dying* from that annually. If a person has been contentedly going to the off-licence for their wine all their lives, they are not going to suddenly start buying it from a website just because the internet has been newly invented. Since an invention is likely to have some element of newness about it, this bodes ill for an inventor.

Here's another example: this one's from the tree surgeons who have just cut down our silver birch. They are running a business that was started by their father and, consequently, are well known in the locality. Recently a rival firm from the neighbouring county announced they were going to expand until they had taken over the whole of the South East of England. They leafleted every household. No result. When folks wanted tree work done they continued to call the number they had always called for the service they knew and trusted.

A good reputation like that doesn't just protect your business from would-be competitors, it can get you new business too. The same tree surgeons now export container loads of English oak logs to the Antipodes! This came about because, for years, they'd been supplying a Sussex company that smoked salmon and then one of that firm's sons emigrated to set up a similar business in Australia.

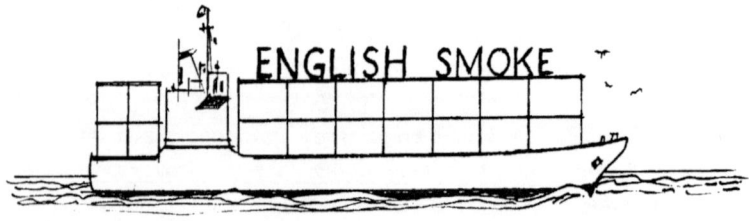

Naturally he wanted the same quality of smoke!

Unfortunately, an invention may be too new or revolutionary to benefit from a tradition or a reputation; perhaps you should sell or licence it to an existing well known brand, or try to get an association with one (see next heading).

Examples of marketing failure are legion and novice businessmen involved in new ventures are especially prone to placing unrealistic faith in the power of advertisement. That failed Education Show I described in the early pages of this book was a new show at a new location run by a new young team of organisers. They obviously were experienced in the industry, probably from working for an established company, and had set up on their own. They had done all the usual publicity tactics: huge posters, pages in the local press, promotion on local radio, direction signs on lampposts, etc, and still hardly anyone came; there was no tradition of a show being there regularly at that time of the year and therefore no reputation of it being worth attending. If only they'd properly established the demand beforehand…

In fact, marketing is sometimes more than just a failure; a badly chosen campaign can actually cause damage to your reputation by alienating potential customers. Take the case of the morning-after pill 'Levonelle'. The manufacturers, Schering Healthcare, had to withdraw a poster campaign using the strap line 'Immaculate *Contra*ception' because of the barrage of complaints to the Advertising Standards Authority from the rabidly religious sector of the public. I still think it was a good play on words but, for Schering, it was just money down the marketing plughole *and* some offended potential customers; although they were probably not much of a loss as those puritanical people were unlikely to indulge in contraception anyway!

Another example features the rock band, Queen. I bet they lost some popularity, and therefore record sales, when they produced that video for 'I Want to Break Free' which showed Freddie Mercury camping it up as a pink aproned housewife sweeping the kitchen floor. Many homophobic former fans must have been driven away. I was lucky enough to *hear* the number years before I saw the video so its greatness was not tainted for me. It's not that I'm homophobic, I just don't think it was an appropriate video to attach to the song; I'd rather have seen a live performance. Give another listen to Brian May's guitar break – how do you make a guitar sound like a better type of synthesiser? I guess you use a lot of the toys found in a recording studio...

You may have heard the old adage: "There's no such thing as bad publicity." **Don't believe it.** Gerald Ratner ruined his high street chain of jewellers, which once employed 27,000 people, with one after dinner speech to the Institute of Directors in which he said that Ratners, *"sold a pair of earrings for under a pound, which is cheaper than a prawn sandwich from Marks & Spencer, but probably wouldn't last as long".* This was shown on the TV evening news and his company collapsed overnight.

If marketing was as good as the marketers would like us to think it is, it could be used to educate the public into spending with confidence on their credit cards over the internet. That should be an easy sell, because it's actually much safer than giving your credit details to someone on the phone, but no-one is running such a campaign. Is it because they know it won't succeed or because no-one has offered to pay the bill? Maybe they are frightened they would be held responsible for any losses...

The marketing world is beginning to recognise that their traditional methods are no longer producing such good results and are trying different strategies. One example of this is the O2 initiative in what was the Millennium Dome. The company is providing an entertainment base which their mobile phone customers will have preferential access to. It's a twenty first century version of 'my gang is better than your gang'.

Having just reread the last few sections, I wonder if I have presented too jaundiced a view of marketing. It's certainly true that it needs to come with a health warning and I readily admit I got my original buzz from the inventing to production side of things (although I'm getting really interested in marketing now – after all, it's just as creative) but, an invention without a market is just a waste of time and money. So what can you do? Bite the bullet and find the money somehow (remember you can spend millions fruitlessly) or, use some rich person's money by selling an interest in the profit from your product to them (they'll probably want as much as 45% of your entire company). You may not live long enough to build up a business based on getting people to visit your website without throwing a fortune at publicity (unless you suddenly become famous); maybe you should think of negotiating with the high street retailers – after all they already have customers browsing in their shops...

It's not that I'm anti-marketing any more than I am anti-retailing. I just want it on my own terms and, hopefully one day, I will get an opportunity to make my own commercials; I already have the screenplay in mind, if not the money to spend in my pocket!

Brand partnerships

One way of getting an instant high quality reputation is to piggyback on an existing brand. A brand is an amazingly powerful thing in marketing. People already know the name and have confidence in the products or services. Just take a look at the Manchester United football team. The soccer players come from all over the world, the manager's a Scot and the owner is an American. There's nothing 'Manchester' about it; it's not a local football team, it's a commercial enterprise with *a brand*. Once you have a label like that you can even sell the name. Just print it on Tee shirts, mugs, pencils, cuddly toys, coasters, flags, anything. It's called merchandising.

On the principle that success breeds success, a fledgling company can try to get reflected glory from an existing brand by an association with them. Your gizmo then becomes the Barbie Gizmo or the Lion King Widget.

Of course the brand owner knows the value of his property and he will make the rules for allowing you to bask in his aura. For a start he won't let his brand be associated with you unless he likes you, or more accurately, thinks your product is suitable for him to get into bed with. That is, he will have to be persuaded that his company's profit line and reputation will both be enhanced by the partnership with you. Slimming pills are unlikely to get an endorsement from Mars bars, or condoms from Care Bears, for example.

Is a Tracie doll right for promoting industrial footwear?

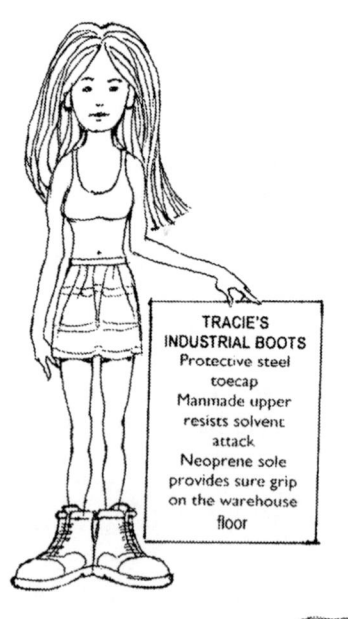

Secondly, he will restrict your access to his logos, comic characters, trade names, etc, to a period of time. At the end of that period all of your unsold product which is identified with him will have to be destroyed so as not to devalue his image by having the price reduced in a sale. Thirdly he will make you pay huge amounts of money for enjoying the privilege of surfing to success on his brand.

Strangely, you don't even have to get associated with a well-known brand. You can simply link your item with an *unbranded product* that folks are already buying and sell it as a package. 'Every cheese sandwich comes with a free Superbo milk-shake!' Charge a good price for the sandwich and you'll get folks to experience your new milk shake product *and* make a profit on the marketing trial. Next week they'll buy the milk-shake on its own if it's any good. You will have benefited from that alliance of the new with the familiar.

Interestingly you can also use this type of association to massage profit margins. In the early days, each unit of your product might cost you so much that there's not enough margin for all the links in the chain between you and the end-user. Your supplier will probably offer you reductions for high volumes but that would mean triggering a threshold requiring you to purchase tens of thousands of pounds worth of stock. Your business might not be able to sustain that amount of risk for a new invention in a new market place. It sounds like a Catch 22, but you could pay for a smaller production run that has a higher unit cost, sell it with an established product that has a comfortable margin and 'share' the profit between the two.

Everyone is doing a variation of this technique. Yesterday I bought a Sunday paper that I normally wouldn't read just because it had a free DVD attached to it. It cost £1.30. I asked what the usual price of that newspaper was. They told me 70p! So much for a 'free' DVD...

Direct Marketing

This is one of the best ways to go with certain products; cut out the middle man and target your end-users. Michael Dell is another of my heroes. His technique of mail shot catalogues to company addresses and leaflets in journals obviously does the business for his range of computer products. He has no distributors, wholesalers or retailers taking their cut from his goods which can be very competitively priced as a result. Dell computers may not have the fastest, most 'state of the art' chips but they are good workaday quality with after sales service agreements and telephone support lines.

The United States has a unique infra-structure for electronic media. Its huge size, recent development and federal nature have meant that *national* terrestrial TV and radio transmissions never took off when they did in Europe. Broadcasting is city based with almost every home on cable. This means that you can treat each city like a separate market and air commercials for little expense. Since there are one hundred cities with populations of 200,000 or more, you have the opportunity to start your marketing campaign small and roll it from city to city making adjustments as needed. You can even ease off transmitting the advertisement for a while if you are generating more demand than you can handle. It's a marketer's dream.

This system is known as DRTV: Direct Response Television. You make a cheap little 30 second commercial mentioning your website address and a call centre number and buy media time with a cable company for one city. You can select the programming it is aired with, so that your intended demographic is likely to view it, and then you sit back and wait for the orders to pour in. If they don't pour in, you can alter parameters like the price, the time of day it's broadcast, the programme it's associated with, or the voice over, and try it in another city. It works phenomenally well when it does work. You need a simple product which solves a common problem like unruly hair or the pasta accidentally disappearing down the plughole when you strain it. The rotating hairbrush and the pasta pot sold millions by this method.

The pasta pot has a few holes drilled in its lid to let the water out but keep the pasta in. You don't have to get the colander out of the cupboard or slide open the ordinary saucepan lid and watch the spaghetti escape. It's not high tech; it didn't need a genius to invent it (years ago we would have gone out into the shed and drilled our own holes in an old saucepan lid) but it touched a nerve. And, after it had exhausted its DRTV promotion, toll free call, credit card purchase potential, it had a secondary market: millions more were sold in 'As seen on TV' stores to folks who like to pay with cash.

Beware though, there are several companies offering to promote your product on TV who encourage you to believe that you will become rich when all that happens is they take your money and the campaign results in no sales. Check out their dissatisfied customers on the internet. It's the same scam that I described some unscrupulous publishers running only slightly higher tech.

When it does work though, advertising really works. You can even sell weekly deliveries of boxes of organic vegetables by direct marketing using simple leaflets printed on recycled paper (they know their market!), and it seems to enhance 'word of mouth'. WoM is the best sort of advertising you can have. If your neighbour, relative, work colleague or friend tells you about a product they've bought, it carries more credibility than any other form of promotion and costs absolutely nothing. Products that are heavily advertised don't attract much WoM; who needs to talk about the fact that they were suckered by an ad? But there is little more natural than chatting over the garden fence and mentioning that you have your organic vegetables delivered...

Marketing is never finished

One thing you need to know about marketing is, like a puppy, it's not just for Christmas. You need to have a continuous programme of publicity or advertising throughout the lifetime of a product. You can step the marketing intensity up and down to vary the demand so that it doesn't outstrip production but you will always need to remind the public of your existence and how they can obtain your products. It even works for milk and eggs which, you might think, would not be affected by marketing...

Drink whisky and eat carrots!

(You'll be able to see where you are when you collapse in the dark!)

RETAILING

High Street Vendors/Supermarkets

As I explained on page 43, some supermarkets and high street retail outlets demand up to a six-fold mark up. Outrageous isn't it. You conceive an idea, patent an embodiment of the gadget, develop a prototype, test market, manufacture and package a product, and they expect to make up to six times as much as you for the trivial effort of taking the customers' money. The tax man also gets more than you! His VAT take is 15% (of the final price) and you are left with 12% while the greedy retailer runs away with 73%! What's going on here? I know shops have rent, business rates, utility bills and wages to pay but this margin is unjustifiable.

Another problem with large retail outlets is their loyalty to existing suppliers. Almost twenty years ago I interviewed Robin Bines who had just introduced Ecover environmentally friendly cleaning products into the UK. He had 2% of the washing powder market at the time. It doesn't sound like much but that represented £10m turnover per annum, which equates to over £20,000,000 a year today! I did a 'he who dares wins' thing. I went to him pretending to be a free-lance journalist, interviewed him and then went to a local magazine and sold them the story. I was actually a supply teacher on a day off but, hey, nothing ventured, nothing gained! He had made a very clever move. Having come up against the resistance of the supermarkets to stocking his brand for fear of offending their two main suppliers (Unilever and Proctor & Gamble) he sidestepped them and got his detergents on the shelves of health food stores. The hippy food and alternative medicine shoppers were more his type of Earth friendly client anyway. After a while the supermarkets *had* to stock his products in order to meet customer expectations. Then he made another clever move and sold the business.

When you are seeking a retail outlet try to think of the most appropriate one. That is, the one where large numbers of your demographic go most frequently. If you are selling a product to mums try to get it in Mothercare. If it's for men then maybe the fishing tackle or motor spares shops. We are attempting to get the Wordwiza into libraries. What better place is there for presenting a reading tool to parents who are helping their children learn to read?

And then there is the problem of simply getting a retailer's attention. Business to business selling can be just as much a communications problem as selling direct to the public. If you approach a chain store's buyer you will find, like I have done, that usually they won't meet you. *'Put a sample in the post'* is the best you will get out of them, which is no help if you are seeking advice with the development of the product or packaging. Sending them a sample puts the ball firmly in their court and leaves you at the mercy of their lack of vision. Some don't even bother to respond. When they do, it may be to tell you that they only sell own-branded goods.

Remember you will probably only get one chance to approach a company and if your product lands on the desk of someone who can't relate to it that will be the end of the matter. It might simply be that he is a hoary old misogynist and your product appeals to girly young women who he doesn't understand and he won't see its relevance. Once he has rejected you, he's not going to alter so you may as well give up on him. It's the same with assessors who award grants and examiners who grant patents; one may respond favourably, another may not but once they've decided you can't persuade them to change their minds. Why would they admit they were wrong the first time?

So it's important to make sure your first contact is a good one. Prepare as well as you can: research the company, make your presentation fit in with their existing business. Get your product as ready for sale as possible and don't expect them to use their imagination about what it might look like in a blister pack with a backing card. Assume they have no imagination. Show them something they could put out for sale today.

After you have been rejected or ignored a few times it's easy to imagine it's the buyer's fault. Maybe they think they're being adventurous when they re-order the same stock as they had last year. Many people do their job in minimalist style; they don't want to initiate something that might generate more work and delay their going home time. If you have employees like that, who have lost enthusiasm for your business, show them the door. Particularly beware of staff who are approaching retirement or leaving to go to another company; they get into 'demob' mode and are just a waste of space.

That is the trouble with *nationalised* industries; there's either no competition to beat, and therefore no need to be economic, or the workers know there's a bottomless pit of funding to pay for unlimited inefficient practices because the government daren't allow the enterprise to fail. The UK post office, having shed 45,000 employees in an attempt to face up to new competition is now suffering from strikes because they need to lose another 40,000! How hopelessly overmanned were they?

Big companies can be treated by their staff in the same way; employees can regard them as indestructible and take all sorts of liberties. When I was a student I did a summer vacation job as a pipe-fitter's mate for a large company which was building a naphtha gas cracking plant. The charge hand used to disappear into the Portaloo with a porno mag for hours! One time I was standing around feeling useless so I said to my pipe-fitter *"Let me have a spanner and I'll undo that nut while you hold the bolt head."* He replied, *"If you start helping we'll have no hours on double pay at the weekend!"*

Getting back to my thread, it's a pity it's so hard to get to see buyers because you can learn so much more in an interview than by correspondence. For example, W H Smith's buyer was good enough to meet me and revealed that, to sell in the UK, pens need to be blue, black or silver whereas on the continent customers are happy with yellow or green. It may be a weird fact about customer expectations, but where else could I have found it out? Unfortunately she then had the effrontery to retire and I could not persuade her replacement to give me a follow up appointment.

As is the case in so many things, it's not what you know but who you know. Take the example of Levi Roots who appeared on Dragons' Den with his Reggae Reggae sauce. He had been making this family recipe chilli sauce and selling it at Notting Hill carnival (the biggest festival in Europe) for ten years. Throughout all of that time he had been trying to get to see the buyers at the big supermarkets without success. Four weeks after featuring on Dragons' Den it was in all branches of Sainsbury's. Dragon Peter Jones simply introduced him to his friend the Sainsbury's buyer! Levi is now opening a Caribbean restaurant in Battersea!

In the recent past, some high street retailers have been able to use their muscle to bully people into accepting their terms but, hopefully, those days are numbered. When enough of the public realise that it's cheaper to buy most things from the internet, then you may see many retail specialists vanish from the high street. The coffee shops, clothes shops and greengrocers will survive because we want to use our senses to experience their products *before* we buy, but how will booksellers, record stores and electrical goods vendors continue to be able to exist? We don't need to pay high street prices for those products. If I was W H Smiths I'd be installing espresso machines. I believe the bookshop concept started life as a coffee house which offered something to read. Perhaps history is about to turn full circle…

Actually, the future is not at all bright for books and bookshops. Don't invest in them. The written word is the next target for an IT solution. When the flexible screen hits the market in 2008 we will all download reading material and hard copies will gradually die out. It will be a repeat of what happened to the music CD when mp3 players became available. You remember? It didn't take long for downloads to become the majority of record sales. Everybody's CD tower turned into a museum exhibit almost overnight. Soon we will carry an iPOD equivalent with a roll up screen and download our novel into it. We might even have the daily paper arrive automatically by wifi subscription. Paper mills may be in for a hard time when this technology arrives. Domestic paper recycling may disappear. What will we put into our wheelie bins?

Don't get me wrong, I do recognise that retailers provide an easy solution to that otherwise intractable problem, marketing, and I *am* prepared to do business with them, (just in case any chain-store's buyer is reading this and would like to stock our Easyriter pen and pencil, or *this book!*) but only under reasonable terms. Both sides have got to be satisfied that they have a fair deal and can work together to make money for each other. That's the only acceptable outcome from a negotiation.

Catalogue companies

Certain kinds of stuff still sells from a glossy catalogue. You can either print your own, which is hideously expensive, or you can get your product into someone else's catalogue. As with retailers, the catalogue companies will come up with their own terms. Some will want your product to appear as though it is theirs. You'll have to get their brand name on it or on the packaging. They may insist on having your goods on sale or return. You are then at their mercy. If they take your product and present it in a pedestrian way so that it doesn't sell, it will have been at your risk. You will have earned nothing and just got a bit older.

We got our prototype reading tool into a huge catalogue which is sent to all schools. It had five hundred pages and it didn't do justice to our product which was given a tiny picture and two lines of text. Fortunately the catalogue company purchased from us so we had a good sale. Whether they resold them or not we shall never know but they didn't re-order. Another catalogue company director turned the reading tool down because he didn't think he could sell it. Perhaps he was right not to stock the product considering their sales strategy, but now the Wordwiza is selling well by word of mouth due to exposure at exhibitions and from our website with its flash demo so I'm glad we don't have to share the income with others. I wonder what that guy thinks when he sees me at exhibitions with a crowd around my stand...

The message is, if your product is not too innovative, exposure in a catalogue may work for you.

MONEY (AND HOW TO GET IT FOR YOUR INVENTIONS)!

Oh, how I wish I knew! Let's face it; an inventor is indulging in an egotistical exercise. It's vanity publishing for products. Most serious inventors end up re-mortgaging their homes. James Dyson was a million pounds in debt at one stage and now he's a sterling billionaire, number 39 in the UK rich list! He is a great advert for dogged determination.

Inventing consumes money like nobody's business. As always there are three ways to finance things (discounting stealing): earn it, beg it or borrow it. I've done all three. I spent a long time at the day job (teaching) and I sent my wives out to work in an effort to pay patent agent's bills, etc. (No, I'm not polygamous, just remarried!) I got a sponsor (Richard) who invested in the business in the expectation that we would make big money one day, and I sold portions of our home to the bank (equity release) to cover debts incurred during the development of our products. None of this was easy. Some people thought we were foolishly wasting money that could have been better deployed on less risky ventures such as buying lottery tickets. They know little about statistics; why is it that people imagine that bad things, like having a miscarriage (about 1 in 4 pregnancies), will never happen to them but good things, like winning the lottery (1 in 14,000,000), will?

I don't do the Lottery, which makes me marginally less likely to win it than the people who do!

Linda Smith English comedian
(Linda died too young)

There are grants though. If you go to your local Enterprise people they will tell you that the government and various non-governmental organisations have schemes providing funds to invest in likely prospects. That may be true it you like filling in forms, keeping meticulous records and hitting targets or going to live in some derelict part of the country which is designated as an 'Enterprise Zone'. So far Ideasun has failed to qualify for much grant money other than support for exhibiting overseas, although we haven't given up trying. If you hire an office services address in another region you can revise the application that failed from your home address and submit it to the bureaucrats there. A different assessor may pass what was previously rejected.

Be warned – most of the funds are available for 'research and development' but not for manufacturing and marketing. *So you need to apply for making your prototype.* Having been rejected a few times by authorities who claim to have millions that they must dispose of before it is clawed back at the end of the financial year, I'm beginning to think the grant system is a mirage. Perhaps the government deliberately allocates funds with unmeetable criteria in order to appear to be supporting British innovation but in reality they would prefer to use the unspent public money to build a Millennium dome or invade Iraq!

Am I too cynical? You judge. Even if you do get one, don't think of a grant as a gift; if your business succeeds, many grants have to be repaid. Not really a grant then... More a sort of loan!

Sources of grants:

www.dti.go.uk/innovate/pdfs/sms_business.pdf
www.is4profit.com/busadvice/grants/index.htm

You may wish to consider forming a limited company and becoming VAT registered. There are pros and cons to both these initiatives. If you're *not* a limited company, and you have another income, you can set expenses against earnings and recover outgoings from personal income tax paid during the current year. Whereas if you become a director of a limited company you will have to wait to recover the money you've invested (your 'director's loans') until you start making a profit. However you will have to submit annual returns to Company House in addition to dealing with your personal tax situation as an employee of your own company which is regarded in law as a separate entity. Confused? Me too!

Being a limited company though can protect you against damages in the event of someone demanding compensation for the harm your product has done to them; you simply declare the company bankrupt and the claimant can only recover the two pound shares that you invested when you set up the company. This may not be moral, but it is commercial. It's not so easy rising from the ashes though, with that on your record...

I'm not wonderful with finance so we became a limited company for a much more silly reason. It was because, when you put Ltd after Ideasun and say it quickly it sounds like 'ideas unlimited'! Ideasun, Ideasun Limited, Ideas unlimited! Ho ho ho! Yes, I know, I'm very easily pleased!

In the States, the equivalent of 'Ltd' is 'Incorporated'. If you're thinking of rushing over there and registering an ink manufacturing company called 'Ink Inc', don't bother. It's already been done!

Value Added Tax registration can be more useful in the early stages. When you are developing the business, any VAT you have paid on purchases for the company you will be able to claim back each quarter. When you start to make sales though, you will have to pay VAT on them. Once your turnover exceeds the threshold you don't have a choice and will have to register for VAT anyway.

If you're like me you will find all this financial stuff very boring and will not want to read about it as you jet to your holiday destination. Just get an accountant.

APPENDICES:

1. *How you can become an Inventor*

Get to know how things work. This is not just about taking machinery apart and examining it, it's about reality and the world; inventing requires an elementary knowledge of Physics and Chemistry, especially the properties of materials. If you don't have this you may waste a lifetime trying to invent perpetual motion and not understand why that project is doomed.

Open up your mind. Retain a childlike wonder and curiosity about everything, especially everything man-made; criticise and query its design, operation and function. Question 'experts' when you get the chance. Challenge orthodoxy. Conventional wisdom can be wrong.

Be interested in everything. Don't be a back row boy who sniggers and misses an opportunity to gain knowledge. Admire information more than entertainment or sport. Get learning – it's a lifetime activity.

Don't worry about being a 'geek' – when your peers jeer at you, just remind them who the richest person in the world is - nerdy Bill Gates! Geeks rule! Be proud to be a 'boff'!

Don't get into a rut doing a job for life. Don't forget, *part-time* employment frees up hours for you to invent... Grasp opportunities! Keep putting yourself into new situations where you have to learn new skills and there is a risk you might fail. Accept a challenge – I bet you can't play the penny whistle!

Think! Spend a lot of time pondering problems but take time off too – solutions often present themselves while you are asleep. That's when our brains organise our thoughts.

Have a go! Attempt to make things, or do things, which might seem crazy – even to you! Don't give up! Have another go! Things rarely come together at the first attempt. If something happens, even if it all breaks or looks like failure at first glance, be ready to learn from it and then apply your new knowledge.

2. The Ten Criteria for an Ideal Product

1. Invent something people *want* not something they need. Any vanity or novelty based product has a good chance of success. Aim to gratify your customers or indulge them.

2. Don't invent something too similar to existing products. If the marketplace is already occupied by strong products, you will have little chance of success. Look for a gap in the market.

3. Don't invent something too new or revolutionary. Any really different new product involves teaching the potential customers what it can do before they will realize that they need it.

4. The most ideal product will appeal to both sexes, all ages and all nationalities.

5. Try to invent something simple and cheap enough to develop at home.

6. You want as big a gap between the cost and the price as possible.

7. The ideal product would be easily packaged and posted from a direct sell, mail order website and/or 0800 telephone fulfilment service in order to eliminate all those middle men who want to hitch a ride on your product and share in the profit margin.

8. To be worth trading, especially if dispatching by post or carrier, there is no point unless the price is within the credit card payment range: at least £1.99 plus p & p but £300 is probably the ceiling.

9. Subscriptions or refill/replacement/accessory purchases make an excellent long term reliable income.

10. Your invention must be something that people want *now*

Warning: A product that meets all ten criteria still gives no guarantee of success; even the best product is prey to the beast of marketing!

3. The Catch 22 Problem

So much of what has to be done to get your invention from a concept to a product ready for the consumer to buy is a 'Catch 22'. For those of you who haven't read Joseph Heller's excellent novel, a catch 22 was when, during the Vietnam War, the US army Doctors wouldn't send a soldier home on grounds of insanity because it was recognised that the combat missions they were flying were so dangerous no sane man would do them. Therefore the very act of requesting to be sent home on grounds of insanity proved that the soldier was in fact sane! The inventor encounters these chicken or egg situations at several stages in the process. Inevitably outgoings will commence before the earnings are coming in to fund the early costs like patent and design registrations. Here are some of the cart before the horse situations you are going to meet:

1. You need to get intellectual protection before you can talk to people who might advise you whether you have a good idea unless (a) you can trust them, or (b) you have a pot of money with which to defend a non-disclosure agreement which they may sign and may yet violate.

2. You must apply for a research and development grant before you have even made a prototype because they won't fund you retrospectively so you have to sit on your hands while the bureaucrats cogitate and your patent ticks away (or lie about what you've done).

3. You have to try to sell your product before you've had it made because you don't want your money locked up in boxes of unsold goods. And it's best to assume chain store buyers are incapable of imagination!

4. You have to commission vast quantities of your product at hideous expense in order to bring the unit price down to a level where you will make a profit and you aren't sure

what it will sell at yet. Alternatively you can order a small amount of product which costs so much per unit you won't be able to sell it profitably!

5. You can't persuade the banks to give you a loan unless you already have masses of equity (e.g. in your own home) or you have already been awarded a grant and/or got a large order. Banks operate very safely – if you really need money they won't let you have any but if you're already well off they ask, "How much do you want?"

Basically, you need to be rich before you can afford to start inventing in order to become rich. Best to choose the right parents or save up before you start inventing.

4. Final Health Warnings

Ok, you've had an idea. Don't get excited. You've just done the easy bit. You've had the concept, now you have to get it to the consumer.

1. Think about it, does it have a big market? I mean a *really* big market. Not just golfers or snooker players, for example, but preferably the entire population of the world.

2. Can you get protection? Your idea needs to be innovative and not something that could be derived from existing patents by a 'man skilled in the arts' (whatever that is). If it isn't protectable you will be giving it away for nothing to would-be competitors from the moment you produce it or even talk about it. The likelihood is they will beat you to the market because they have the manufacturing and/or vending capability which you lack.

3. Can you afford it? Developing your idea will be ferociously expensive. Expect to get into serious debt. You'd better have a big asset to mortgage first, or a sugar daddy/mommy.

4. Maybe you've succeeded in getting intellectual protection, don't start frittering away the millions you imagine your idea will earn; all that has happened is it's now begun to cost you and you still have to get it manufactured.

5. Well done! You've got it manufactured. Let's face it, that's not too difficult if you can afford to pay someone to make it for you. I hope you've succeeded in getting it made incredibly cheaply. Now your problems are really going to start.

6. Marketing. This is the big one. How are you going to spread the word? Even the best new product requires marketing. Establishing a route to the consumer is nearly

impossible. Retailers will turn you away saying they develop their own products in-house or expect you to provide a display and advertising for them! Marketing costs prodigious amounts of money. However much you have spent on your project up to now was peanuts. Most companies allocate ninety per cent of their costs to their marketing budget – that's how much it normally takes. Better get used to seeing those boxes of product hanging around...

Why don't you give serious consideration to going on Xfactor instead!

Good luck!

5. Those Formative Years
The 'a-sibling effect' and other factors

> *Give me a boy of seven and I'll show you the man.*
>
> **Jesuit saying**

Sometimes I get asked what my motivation is. Since the questioner obviously doesn't understand me, the implication is that I'm not normal! Well, if normal consists of being fanatical about a football team and getting fighting drunk, then I'm happy to admit that I'm not. But I do like (in no particular order!) humour, cricket, women, Top Gear, wine and beer (in moderation) so I only feel a little bit weird.

Early on in this book I referred to the late Professor Eric Laithwaite's impression that being an only child may have influenced his creativity. I call this 'the a-sibling effect' (you might remember from your Biology lessons that an 'a' prefix means 'without') and I didn't dwell on it at the time because it wasn't the main thread of my narrative then, but I can develop the theme now.

I was a child on the Isle of Wight in the nineteen fifties. It was a great time and place for a childhood. Being an only child meant that I could wake up each morning free from the need to socially interact with a brother or sister. I've watched siblings in action; in fact, I've raised a couple. They love to annoy each other just to provoke a response. If they don't get a big enough response they wail, *"He won't play with me!"* What a waste of days. Maybe it was the absence of a time consuming close relationship that enabled me to become fascinated with information. I made a good student and always came top of the class in all subjects at primary school except on the occasions I was beaten by Jane Dale when I came second! I won the races too and came to believe in the myth that I could do anything!

Unfortunately modern society is not geared up to accept or understand people who are generalists. Everyone tries to pigeonhole and stifle us. When I told my financial advisor that I'd written this book he said something along the lines of I should keep quiet about it because investors like a person to be focused on the business. Later he hinted that he considered me to be an inventor and, therefore, maybe I couldn't do entrepreneurship! He had me firmly stereotyped! In case he's reading this, I see him as a mere office worker!

While my childhood mates were jabbing their sisters with an elbow or play-fighting their brothers, I was dismantling an old piano and making things from the components in my father's well-equipped workshop. I made a catamaran out of two of the piano keys, using a dowel from one of the hammers as a mast, and I tested it on the garden pond. Then I used the cylindrical lead weights, which had counterbalanced the keys, as bullets for my homemade gun. Unfortunately I leant the gun up against the side of the wooden garden shed and the recoil punched a hole through the wall! Practical experience of the Law of Action and Reaction!

We didn't have much in post-war Britain, no Playstations or Xboxes and not even television until I was eleven, so there was an acceptance that probably the only way to get something you wanted was to make it. Several of my peers at the Grammar school made quite serviceable electric guitars or built transistor radios. Others flew model aircraft constructed from balsa wood and paper treated with dope (No, not marijuana. 'Dope' was the name for a type of varnish which tightened and strengthened the tissue paper used to cover the wings and fuselage). I made a 'harmonograph' from Meccano, which used the newly available ball pens to draw geometrical designs. As mentioned before, later in life I did all sorts of ambitious DIY on our home.

I had piano lessons when I was seven and, during my time as a teacher, I taught myself the electric guitar and played and sang in a band. That was fun! I also played bass for a Trad Jazz band and joined in the school orchestra for performances of Oliver! (bass) and Godspell (lead guitar). The recent research on the 'Mozart Effect' (see New Scientist) indicates that all of this may have been intellectually beneficial.

My father had been paying into a little insurance policy for me which matured when I was sixteen and enabled me to buy a second-hand motor scooter. I spent many enjoyable hours tuning the two-stroke single cylinder engine by advancing and polishing the ports etc. As you'll have gathered if you've read this far, my father was very supportive of my tinkering and even set up cottage industry production lines for some of my products later in life.

What I am leading up to is the hypothesis that my early upbringing might have influenced my later development. It's certainly possible but I don't think my circumstances were very remarkable. Lots of children were having childhoods like mine in those days. Today attitudes are different.

Now, if you want something to fill up your leisure time, you go out and buy the latest radio-controlled helicopter or a laptop so you can play Sims. It isn't necessary to develop either ingenuity in scavenging materials or any manual skills, other than operating a game controller. We have become consumers dependent upon a leisure product industry. We no longer make our own games and toys; we play 'Spiderman' video game because that is what the manufacturers offer us.

When I want to do a fogey rant, I say to young people, *"You are so spoiled today with your Xboxes, MP3 players and mobile phones. All I had to play with was a stick!"* It's more or less true – we'd cut a stick from the bushes (boys were allowed to carry pen knives then) and beat a path through the stinging nettles to make a secret camp. Putting that ancient history aside, the modern-day commercialisation of enjoyment has somewhat curtailed individual innovation. Fortunately it's not all-pervasive; our youngest got less fun from the toy than he did from playing with the box it came in.

If there is an a-sibling effect it is probably due to freedom of opportunity. Singletons have the luxury of time to themselves to fill how they wish, whereas close brothers and sisters have a relationship to waste their days on. But you don't have to be an only child to get that freedom: a single boy in a sisterhood of girls or the latecomer in a large family may be able to indulge themselves in their own pursuits too. Thomas Edison was the final child of seven siblings, Richard Arkwright the youngest of thirteen, Charles Parsons the sixth son out of six, Trevithick was also the last of six and my own father was an 'after thought' in his large family. That list seems to indicate that perhaps there's a link with being the last born in the family. Maybe having considerably older and therefore more experienced brothers and sisters to learn from gives the youngest a boost. This would support the singleton advantage theory since his home companions are his parents - full grown adults. Then again, yesterday I heard on the radio about research indicating that the elder sibling matures earlier and develops two extra IQ points due to the beneficial effect of tutoring his juniors. So you pays your money and you takes your choice. Alternatively you might arrange to be brought up as a solo by your grandparents, as happened to Leonardo da Vinci. Anyway, someone needs to do some proper research on this matter.

As for motivation, what puzzles me is why others *don't* have it. I can't understand why some people get bored. We've all got versions of the best piece of equipment ever produced: the human brain. Use it or lose it! My brain is the type that doesn't see the point of doing crosswords, jigsaws or the Rubik's cube. To me that's like the Princesses of olden times who spent their days unravelling balls of string that their handmaidens had knotted for them. My late mother-in-law, who was a wonderful grandmother, used to while away hours, surrounded by clutter, playing Patience (Solitaire); eventually she developed Alzheimer's disease. Personally, I prefer a more original challenge than solving puzzles which were created by someone else for no better purpose than simply passing the time; in other words making ageing slightly less painful. Therefore I always have several projects on the go with more at the thinking-about stage.

Certainly I was privileged as a child and came to expect to be able to do my own thing, which makes me a very frustrated employee or holiday companion today. I don't like working for other people because it takes me away from my own projects and we now try to have our annual holiday over the Christmas and New Year period. This stops me tearing my hair out because England is closed for a whole fortnight and it distracts me with some interestingly different location.

Postscript

"Know your stuff: creativity requires expertise; but don't know it too well: overspecialisation puts blinders on. Imagine the impossible: many breakthrough ideas at first seem outright crazy; but you have to be able to impose your idea: crazy ideas remain crazy if they cannot survive critical evaluation. Finally, be persistent: big problems are seldom solved on the first try, or the second, or the third; but remember to take a break: you may be barking up the wrong tree, so incubate a bit to get a fresh start."

Dean Simonton, Professor of Psychology, University of California, Davis

Finally, just remember: the journey from concept to the consumer may involve actual travelling to gain acceptance as an inventor and to avoid being regarded as a prophet in your own land.

The Last Word

At the beginning of this book I referred to our tendency to regard inventing as 'a shortcut to wealth involving minimal effort.' I hope I have put you straight on that one. It's more like trying to scrabble up a vertical glass cliff using only your fingernails!

Printed in the United Kingdom
by Lightning Source UK Ltd.
128413UK00001B/268-471/A